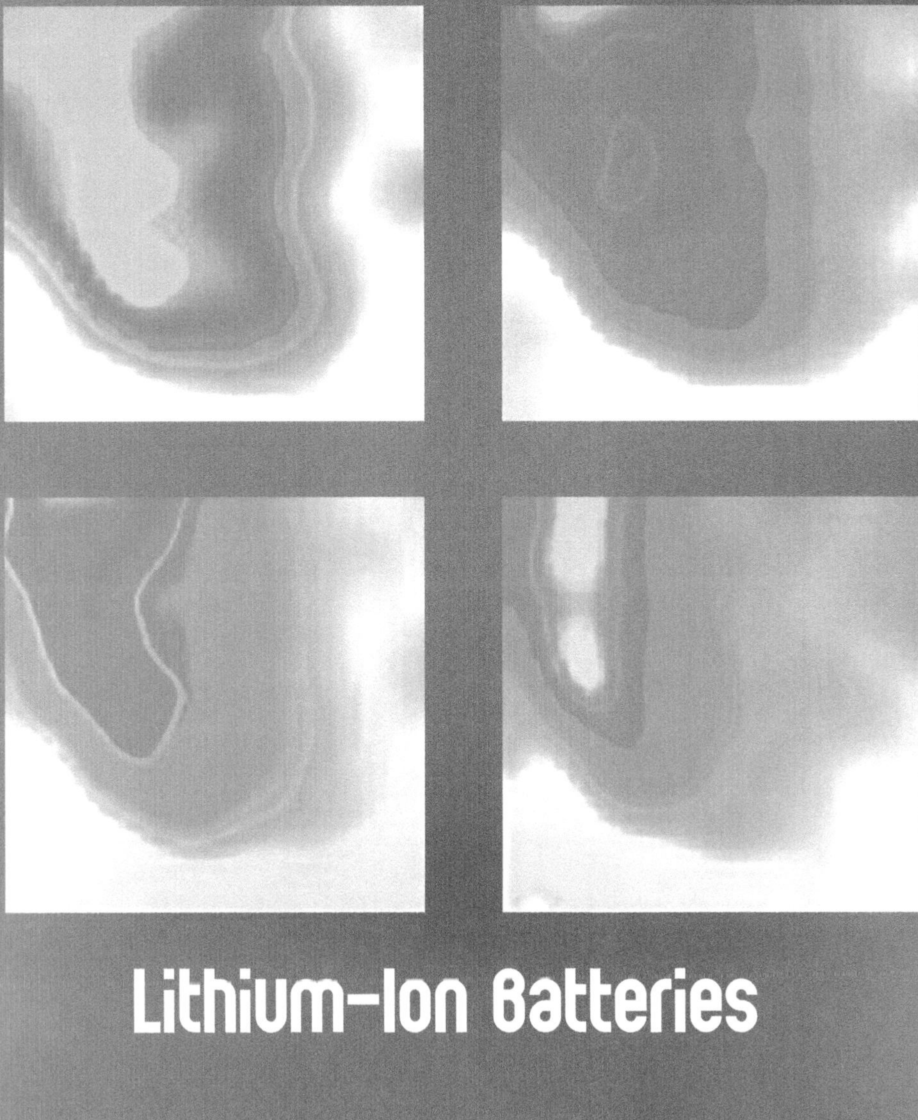

Lithium-Ion Batteries

Lithium-Ion Batteries

Overview, Simulation, and Diagnostics

edited by

Yoshiaki Kato

Zenpachi Ogumi

José Manuel Perlado Martín

PAN STANFORD PUBLISHING

Published by

Pan Stanford Publishing Pte. Ltd.
Penthouse Level, Suntec Tower 3
8 Temasek Boulevard
Singapore 038988

Email: editorial@panstanford.com
Web: www.panstanford.com

British Library Cataloguing-in-Publication Data
A catalogue record for this book is available from the British Library.

Lithium-Ion Batteries: Overview, Simulation, and Diagnostics

ISBN 978-981-4800-40-2 (Hardcover)
ISBN 978-0-429-25934-0 (eBook)

Contents

Preface

Secondary batteries, also called rechargeable batteries or storage batteries, are used in many devices, ranging from portable electronics to automobiles. Secondary batteries are the key component in energy security and the realization of a low-carbon and resilient society. For example, high-performance secondary batteries are indispensable for electric automobiles, power storage of renewable energies, load leveling of electric power lines, base stations for mobile phones, and emergency power supply in hospitals.

Extensive efforts are being undertaken to develop next-generation secondary batteries, with higher performance in terms of energy density, specific power, charge/discharge efficiency, operating voltage, cycle durability, safety, and reduced cost. A detailed understanding of various phenomena that take place in secondary batteries is required for developing higher-performance batteries.

Among various types of secondary batteries, a lithium-ion battery is most widely used due to its high energy density, small memory effect, and low self-discharge rate. This book is intended for students, young researchers, and engineers as an introduction to lithium-ion batteries, with emphasis on diagnostics and simulation of lithium ions in the batteries. Chapter 1 provides an overview of lithium-ion battery, covering principles, structures, and materials and the electrochemical reactions in the electrodes and electrolytes that make up typical liquid-type lithium-ion batteries. Chapter 2 describes advanced lithium and lithium-ion batteries and includes the roadmaps for developing advanced secondary batteries. A part of this chapter is devoted to all-solid-state lithium-ion batteries, which are intensively developed for electric vehicles and other applications due to their inherent safety and high energy density.

Chapters 3, 4, and 5 are devoted to the diagnostics of lithium-ion batteries. Chapter 3 describes advanced diagnostics, mainly with synchrotron X-ray radiation, enabling diagnostics of the lithium-ion dynamics under operation of the batteries. Chapters 4 and 5 describe new diagnostic methods based on ion beam analysis techniques to measure mesoscopic distribution of lithium ions in battery electrodes. Chapter 5 describes the application of this method in the diagnostics of lithium-ion batteries, showing lithium-ion distribution under various operating conditions.

Chapter 6 provides a simulation of the lithium-ion battery, where the transport of lithium ions and electrons is analyzed on the basis of the equations describing the electrochemical processes, the lithium-ion motion, and electron conduction. Chapters 7 and 8 are devoted to future prospects of the advanced diagnostics of lithium-ion batteries. Chapter 7 describes the development of a compact ion beam source, generated with a compact intense laser, which could be used for ion beam diagnostics of lithium-ion batteries. Chapter 8 describes the prospect of in situ diagnostics of lithium-ion batteries using various approaches, including ion beam analysis, Raman spectroscopy, and neutron scattering.

Some of the contents described in this book were obtained under the programs mentioned below. We would like to thank NEDO, JST, MINECO, FP7, Kyoto University, and QST for operating these programs.

1. NEDO Research and Development Initiative for Scientific Innovation of New Generation Batteries (RISING Battery Project, FY 2009-2015)
2. JST-MINECO Strategic Japanese-Spanish Cooperative Program on Characterization of Advanced Electrode Materials by Means of Ion Beam Analysis Technique for Next Generation Li-ion Batteries (FY 2011-2014)
3. JST CONCERT-JAPAN Program within the framework of the Strategic International Research Cooperative Program (SICP) on "Understanding Mesoscopic Behaviour of Li ion in All-solid-battery (UMBLA)," coordinated with ERA-NET project funded by the 7th Framework Program (FP7) of EU Research Funding 2013-2014
4. Collaborated Research Program of the Institute for Chemical Research, Kyoto University

5. Facility Users Program, Takasaki Advanced Radiation Research Institute, Quantum Beam Science Research Directorate, National Institutes for Quantum and Radiological Science and Technology (QST)

We are also indebted to Toyota Central R&D Labs., Inc., for participation in programs 1, 2, and 5 and Toyota Motor Corporation for participation in program 3. We would like to thank all the authors of this book for writing manuscripts with updated content. In addition, we would like to thank the staff of Pan Stanford Publishing for their patience toward completion of this book.

Yoshiaki Kato
Zenpachi Ogumi
José Manuel Perlado Martín
2019

Chapter 1

Overview of Lithium-Ion Batteries

Yuki Orikasa[a] **and Yoshiharu Uchimoto**[b]
[a]*Department of Applied Chemistry, Ritsumeikan University,*
1-1-1 Nojihigashi, Kusatsu, Shiga 525-8577, Japan
[b]*Graduate School of Human and Environmental Studies, Kyoto University,*
Yoshidahonmatsucho, Sakyoku, Kyoto, Kyoto Prefecture 606-8501, Japan
orikasa@fc.ritsumei.ac.jp

1.1 Introduction

1.1.1 Batteries

The word "battery" generally refers to a chemical battery, which means any device that extracts electrical energy from chemical energy. Among the many chemical batteries available, primary batteries extract the energy once via an electrochemical reaction and cannot extract any more energy after the active material has reacted completely. Here, the "active material" means the electrode material that has the chemical energy. On the other hand, secondary batteries (or rechargeable batteries) allow the reverse reaction to occur when electrical energy is supplied, even after the active material has reacted completely, to regenerate the active material. In other

Lithium-Ion Batteries: Overview, Simulation, and Diagnostics
Edited by Yoshiaki Kato, Zenpachi Ogumi, and José Manuel Perlado Martin

ISBN 978-981-4800-40-2 (Hardback), 978-0-429-25934-0 (eBook)
www.panstanford.com

words, they can "store" electrical energy in the form of chemical energy, and in that sense, a secondary battery is also called a storage battery. A fuel cell is a battery that can extract electrical energy from a fuel, such as hydrogen as the negative electrode–active material, using oxygen from the air as the positive electrode–active material. This is the ultimate type of battery because there is no theoretical limit to the amount of energy that can be extracted. However, many problems still remain to be solved before it can be used in practical applications.

Secondary batteries have a long history, with lead-acid batteries being used since the 1800s. After more than a hundred years, they took the form of today's lithium-ion secondary batteries (called lithium-ion batteries [LIBs]), and new types of secondary batteries are expected in the future, taking efficiency and the environment into account.

The following is a brief description of the terminology used to discuss the battery characteristics:

Voltage: Corresponds to the difference between the potential of the positive electrode and the negative electrode, which is expressed in units of volts (V). This usually is expressed as the potential relative to the negative electrode.

Capacity: Shows the amount of electricity that can be extracted from a battery. The amount of electricity is usually expressed in coulombs (C), but ampere-hours (Ah) is commonly used to evaluate the characteristics of batteries and electrodes. The capacity is often expressed as the capacity density per unit mass or per unit volume.

Rate characteristics: Corresponds to the "speed" with which charging and discharging are performed, and the watt (W) is used as the unit. To compare capacity retention at different discharge speeds and of batteries with different capacities, the so-called C-rate is used. It is defined as the capacity divided by the discharge time, and a rate of 1 C corresponds to the full capacity being discharged within 1 h. As further examples, a 2 C discharge rate would mean discharging the full capacity within 0.5 h and C/10 would correspond to a 10 h discharge.

Energy density: Indicates the amount of energy that can be extracted per unit mass, expressed in units of Wh/g or Wh/L. This corresponds to the multiplication of the voltage and the capacity density.

Cycle characteristics: This is the capacity retention characteristic after repeated charging and discharging.

A battery is formed from an oxidizing agent (positive electrode), a reductant (negative electrode), and an electrolyte that conducts ions. The energy can be expressed as follows:

$$\text{Energy (Wh)} = \text{Capacity (Ah)} \times \text{Voltage (V)}$$

Therefore, batteries with high operating voltages are desired.

1.1.2 Lithium-Ion Batteries

LIBs are used for electronic equipment and many other applications because they have a range of advantages, such as a high operating voltage, small self-discharge, and no memory effects. As the technology in today's society becomes increasingly mobile and as information networks grow, the importance of LIBs has increased considerably. Recently, amidst the increasing concern for the environment and energy problems, many attempts have been made to install LIBs in plug-in hybrid vehicles or electric vehicles or to use them in next-generation transmission networks known as smart grids. For these reasons, throughout the world attention has been focused on secondary batteries, particularly LIBs. In addition, the demand for improvements in battery characteristics is growing exponentially. As shown in Table 1.1, even today, LIBs are overshadowing batteries of prior art in terms of the energy density. However, for LIBs to establish a solid position in the future, such as being used as the power sources in cars, further improvements will be needed with regard to the output characteristic, cycle characteristic, and safety.

Table 1.1 Energy density of major commercial secondary batteries

		Energy density			
		Wh/kg		Wh/L	
Battery type	Commercialization year	Practical	Theoretical	Practical	Theoretical
Pb/PbO_2	1859	30–50	161	50–100	720
Cd/NiOOH	1899	65	209	210	751
$LnNi_5H_6$/NiOOH	1990	90	275	340	1134
$LiC_6/LiCoO_2$	1991	170	360	460	1365

1.1.3 History

Research into lithium primary batteries, using metallic lithium as the negative electrode, began in the 1960s. Lithium has the lowest potential (−3.04 V vs. a standard hydrogen electrode [SHE]) of all metals, and it is the lightest (ρ = 6.94 g/mol). Therefore, it has attracted attention as a negative electrode for which a high energy density can be obtained. The superiority of metallic lithium for use in negative electrodes was first reported at the beginning of the 1970s [1]. In 1972, Exxon reported a "lithium secondary battery," where Li^+ can be inserted and extracted in the positive electrode and Li metal can strip and plate as the negative electrode. The reaction can be charged and discharged repeatedly, as expressed in Eq. 1.1 [2]:

$$\begin{aligned} &\text{Positive electrode:} \quad x\mathrm{Li}^+ + \mathrm{TiS_2} + xe^- \Leftrightarrow \mathrm{Li}_x\mathrm{TiS_2} \\ &\text{Negative electrode:} \quad x\mathrm{Li} \Leftrightarrow x\mathrm{Li}^+ + xe^- \end{aligned} \tag{1.1}$$

This idea attracted attention throughout the world but did not achieve practical application. The great hindrance to its application was that irreversibility of the reaction would arise at the metallic lithium negative electrode. The problem is that as charging and discharging are performed repeatedly at the lithium negative electrode, lithium is not deposited homogeneously over the lithium negative electrode but forms branched shapes called dendrites. This inhibits the proper cycle characteristics, and as the dendrites grow and penetrate the separator, they come in contact with the positive electrode, resulting in shorts, firing, and potential explosions. In addition, the high reactivity of the lithium metal itself is a problem from the viewpoint of safety.

However, the idea of using an electrode, such as TiS_2 with Li^+ inserted and extracted, has attracted a great deal of attention, and in 1981, Sanyo Electric reported the use of graphite carbon as the negative electrode with Li^+ inserted and extracted. From this discovery, the idea of a "lithium-ion secondary battery" that could perform charging and discharging just by the movement of Li^+, without requiring the passage of metallic lithium, has attracted considerable interest (Fig. 1.1). The form that is mainly used today, using a carbon material as the negative electrode and a lithium-containing transition metal oxide, such as $LiCoO_2$, as the positive electrode [3]. Sony commercialized the first LIB in 1991.

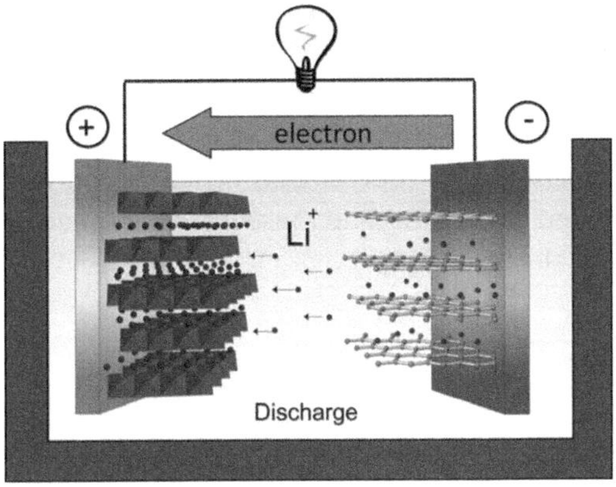

Figure 1.1 Schematic figure of a LIB.

1.1.4 Principle and Structure

The following provides a summary of the operating principles of a LIB. For LIBs that are currently on the market, a lithium-containing transition metal oxide, such as $LiCoO_2$, is used as the positive electrode, graphite is used as the negative electrode, and an organic electrolyte, in which a lithium salt is dissolved, is used as the electrolyte. The charging and discharging reactions of this battery can be expressed as follows:

$$
\begin{aligned}
&\text{Positive electrode:} && Li_{1-x}CoO_2 + xLi^+ + xe^- \Leftrightarrow LiCoO_2 \\
&\text{Negative electrode:} && Li_xC_6 \Leftrightarrow xLi^+ + xe^- + C_6 \qquad (1.2) \\
&\text{Total:} && Li_{1-x}CoO_2 + Li_xC_6 \Leftrightarrow LiCoO_2 + C_6
\end{aligned}
$$

Charging causes the Li^+ within the crystal structure of the positive electrode material to be extracted and dissolved in the electrolyte, whereas the Li^+ in the electrolyte is inserted into the carbon of the negative electrode material. To maintain electrical neutrality, the electrons move through an external circuit from the positive electrode to the negative electrode. The opposite reactions occur during discharge, where the lithium ions and electrons move from the negative electrode to the positive electrode. Both the positive and negative electrodes allow Li^+ insertion (intercalation), and Li^+ insertion and extraction can occur without major changes to the host structure of the electrode. Therefore, a battery with excellent charging/discharging reversibility can be achieved.

The configuration of a representative LIB is shown in Fig. 1.2. Generally, a separator containing an electrolyte is sandwiched between a positive electrode coated on an Al foil and a negative electrode coated on a Cu foil, and these layers are piled up to increase the volume of electrodes [4].

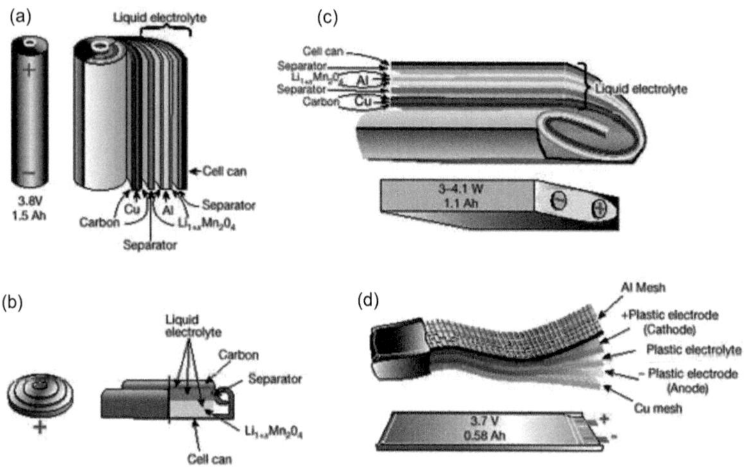

Figure 1.2 Schematic drawing showing the shape and components of various Li-ion battery configurations (a, cylindrical; b, coin; c, prismatic; and d, thin and flat). Reprinted by permission from Macmillan Publishers Ltd: *Nature* (Ref. [4]), copyright (2001).

Generally, the charge/discharge behavior is represented by the electric capacity on the horizontal axis and the voltage on the vertical axis. Figure 1.3 shows the charge/discharge profiles of $LiCoO_2$/Li and graphite/Li cells. In the absence of a side reaction, the electric capacity corresponds to the amount of extracted/inserted lithium ions in the electrode, so the horizontal axis represents the lithium composition of the electrode [5]. On the other hand, the voltage on the vertical axis is the potential difference between the positive and negative electrodes, which corresponds to the energy. Figure 1.4 shows the potential profiles of the transition metal oxides, but the potential varies greatly depending on the active material, and when the positive electrode with the high potential and the negative electrode with the low potential are combined, the battery voltage increases, which realizes the high-energy-density battery [5].

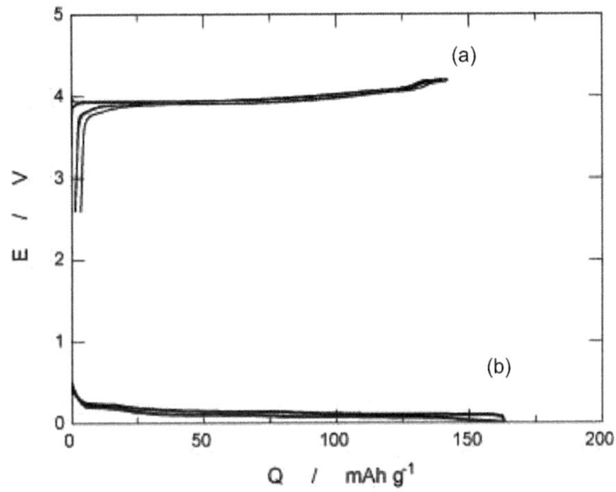

Figure 1.3 Charge and discharge curves of (a) Li/$LiCoO_2$ and (b) Li/graphite cells. The combination of $LiCoO_2$ and graphite gives a lithium-ion battery. The specific capacity of graphite (b) is given in 1/2 reduction. Reprinted from Ref. [5], Copyright (2007), with permission from Elsevier.

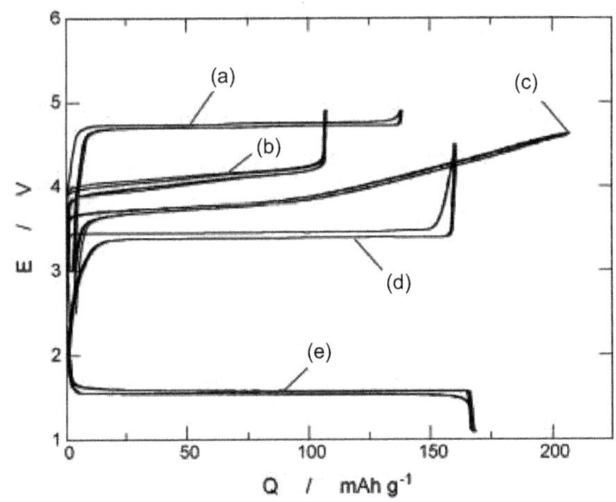

Figure 1.4 Charge and discharge curves of (a) $Li[Ni_{1/2}Mn_{3/2}]O_4$, (b) $LiMn_2O_4$-based material of lithium aluminum manganese oxide (LAMO), (c) $LiCo_{1/3}Ni_{1/3}Mn_{1/3}O_2$, (d) $LiFePO_4$, and (e) $Li[Li_{1/3}Ti_{5/3}]O_4$ examined in nonaqueous lithium cells. Reprinted from Ref. [5], Copyright (2007), with permission from Elsevier.

1.2 Positive Electrodes

The positive electrode–active material functions within the battery as an oxidant during discharge. The main targets of research during the early studies of the positive electrode–active materials in lithium secondary batteries were based on an intercalation reaction centered on transition metal chalcogenides, such as TiS_2 [2], which has a 2D layered structure, and $NbSe_3$ [6], which has a 3D chain structure. American companies, such as Exxon and Bell Laboratories, have conducted active research in this area. Chalcogenides, such as sulfides, exhibit comparatively good electronic conductivity due to covalent bonding, and van der Waals gaps, which become the sites for the intercalation insertion of lithium in the low-dimensional structure. Therefore, these properties are believed to be useful for high lithium-ion diffusivity within the electrode at room temperature.

After that, Goodenough's group reported layered rock-salt-type oxides [7, 8] in which lithium was contained, as well as spinel oxides [9], as the positive electrode materials of the 4 V-class, which caused the main line for positive electrode research to move toward oxide systems. The reasons for this included the achievement of higher capacities due to the low molecular weight and higher voltages because of the increase in electronegativity. These discoveries by Goodenough made a great contribution to the applications of LIBs.

However, the first-generation 4 V–class positive active electrode materials, namely $LiCoO_2$, $LiNiO_2$, and $LiMn_2O_4$, in the three types of lithium-ion second batteries all use rare metals. In addition, a high voltage is realized using a trivalent/tetravalent redox reaction involving the Co^{4+} or Ni^{4+} states, which are chemically unstable, or a Mn^{3+} high-spin state, which is a Jahn–Teller unstable ion. From the viewpoints of stability and economy, these have become a bottleneck toward practical applications of secondary batteries in electric cars and of large-scale storage batteries. The next-generation electrode-active materials and next-generation batteries are expected to be produced using more common metals as the electrode material.

1.2.1 Layered Rock-Salt-Type Oxide Electrodes

$LiCoO_2$, which has a layered rock-salt-type (α-$NaFeO_2$) structure (space group *R*-3*m*), is used mainly for the positive electrodes of

LIBs. $LiNiO_2$ is also known as a positive electrode material and has a similar structure. In these materials, a 2D plane forms, in which the transition metal and the lithium are arranged regularly in the <111> direction of the cubic rock-salt-type structure, and Li^+ lies between the layers formed by MO_6 (M = transition metal) octahedra (3*a* sites). The transition metal ions and O^{2-} occupy the 3*b* sites and 6*c* sites, respectively. Within this crystal, as shown in Fig. 1.5, Li^+ has a diffusion path on the 2D planes between the MO_6 layers. As shown in the figure, there is nothing other than Li^+ on this diffusion path. Therefore, the diffusion coefficient has a relatively high value (for $LiCoO_2$ it is $D_{Li} = 10^{-11.6}$ cm^2/s) [10].

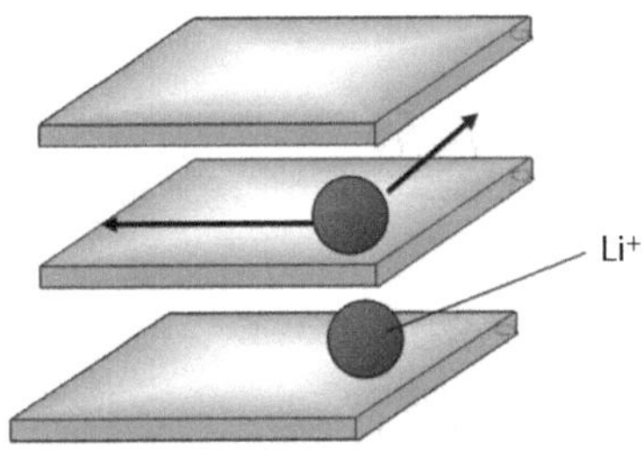

Figure 1.5 Schematic image of Li^+ diffusion pass of an α-$NaFeO_2$-type transition-metal oxide.

The use of $LiCoO_2$ as a positive electrode–active material began in 1980 when Mizushima et al. discovered that $LiMO_2$ compounds have a potential of 4 V relative to metallic lithium [7]. $LiCoO_2$ is a black-colored oxide with a hexagonal crystal structure, as shown in Fig. 1.6. Its synthesis is extremely simple, and it can be obtained by mixing a lithium compound (lithium carbonate, lithium hydroxide, etc.) with a cobalt compound (cobalt hydroxide, cobalt carbonate, etc.) and heating it in air [11]. In addition, even if the starting material is changed, a large difference in the $LiCoO_2$ characteristics is not observed [12]. $LiCoO_2$ has a theoretical capacity of 274 mAh/g assuming that all the Li^+ is extracted. In reality, however, the range for which Li^+ can be inserted into and extracted from Li_xCoO_2 reversibly is limited to $0.5 < x < 1$, so the practically extracted capacity is approximately half the theoretical value [13, 14]. In Li_xCoO_2, it has been indicated that if the amount of Li^+ that is extracted reaches 0.4 to 0.5 then a phase transition occurs (hexagonal $\Leftrightarrow$ monoclinic) [15], so the crystal structure cannot be maintained [16]. The dissolution of Co^{+4} due to excessive Li^+ extraction has also been reported [17], and there has been considerable research into suppressing this phenomenon using a metal oxide coating [18].

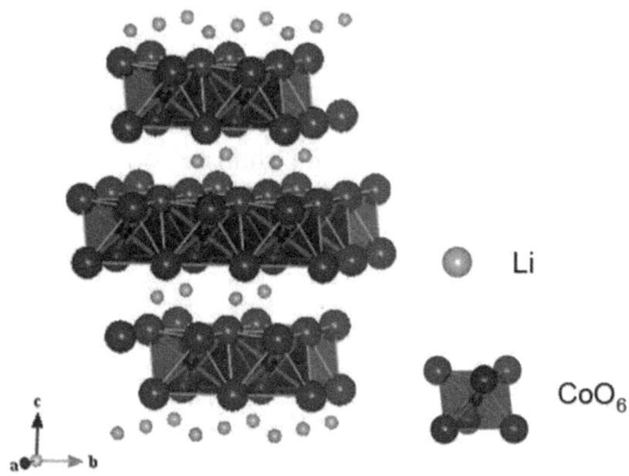

Figure 1.6 Crystal structure of $LiCoO_2$.

Significant research has focused on other layered positive electrode materials, in which the Co or Ni of $LiCoO_2$ and $LiNiO_2$ is replaced by another transition metal. Examples that can be given are $LiNi_{1/3}Co_{1/3}Mn_{1/3}O_2$ and $LiNi_{0.5}Mn_{0.5}O_2$, as reported by Ohzuku et al. [19–21]. In contrast to $LiNiO_2$, in which nickel exists with a valence of +3, the valences of Ni, Co, and Mn in these materials exist in the +2, +3, and +4 states, respectively. The Mn^{4+} in these substituted materials cannot be oxidized further, so it is electrochemically inactive. Therefore, it tends to enhance the thermal stability of the electrode in the highly oxidized state [22]. As Mn^{4+} is inactive, the charge compensation with Li^+ extraction from $LiNi_{0.5}Mn_{0.5}O_2$ is achieved by nickel (Ni^{2+}/Ni^{4+}) or that from $LiNi_{1/3}Co_{1/3}Mn_{1/3}O_2$ by nickel (Ni^{2+}/Ni^{4+}) and cobalt (Co^{3+}/Co^{4+}). In these materials, the amount of the rare metal Co is less than that in $LiCoO_2$. Therefore, these materials are expected to be next positive electrode materials that can replace $LiCoO_2$. However, compared to $LiCoO_2$, Li^+ diffusion is slower, so the rate characteristics are lower. In particular, for $LiNi_{0.5}Mn_{0.5}O_2$, the exchange of Li^+ and Ni ions occurs between the 3*a* sites and 3*b* sites. This is similar to what happens in $LiNiO_2$, which reduces the rate and cycling characteristics. Problems have also been indicated for $LiNi_{1/3}Co_{1/3}Mn_{1/3}O_2$, for example, capacity deterioration is pronounced if it is charged to a potential above 4.5 V (vs. Li) [23].

1.2.2 Spinel-Type Positive Electrodes

$LiMn_2O_4$, a positive electrode material with a spinel-type structure, has a theoretical capacity of 148 mA/g, which is less than that of $LiCoO_2$ and $LiNiO_2$. However, Mn is inexpensive and abundant, has a relatively low environmental impact, and has high safety during charging. In addition, as in the case of the layered rock-salt-type oxide, the structure and the physical properties are closely related. Therefore, it has received particular attention as a material for research and applications. In $LiMn_2O_4$, in the cubic space group *Fd*-3*m*, the 8*a*, 16*d*, and 32*e* sites are occupied by Li^+, Mn ion, and O, respectively, whereas the 16*c* site corresponds to an interstitial cavity (Fig. 1.7a). The vacancy at this *16c* site plays an important role in Li^+ diffusion. In contrast to the positive electrode with a layered structure, the spinel-type has a 3D Li^+ diffusion path, the advance being along a zigzag path, such as 8*a* → 16*c* → 8*a* through the empty 16*c* site that is alongside the 8*a* site (Fig 1.7b) [24]. In addition to the theoretical capacity, another problem with $LiMn_2O_4$ is that the cycling characteristics are low. This can be attributed to distortion of the crystal structure due to the dissolution of Mn ions to the electrolyte and the existence of the Jahn–Teller ion, Mn^{3+} [25, 26]. Therefore, several studies have been performed to improve the charging/discharging characteristics by replacing some of the Mn with a different metallic element [27]. In addition, from the viewpoint of achieving even higher energy, spinel oxides that are substituted with Co, Cr, Ni, and Fe, which operate at high potentials over 4.5 V, have received attention as the so-called "5 V–class positive electrodes" [28, 29]. In particular, those substituted with Ni provide high capacity and excellent charging/discharging characteristics and are expected to be electrode materials for high-energy-density batteries.

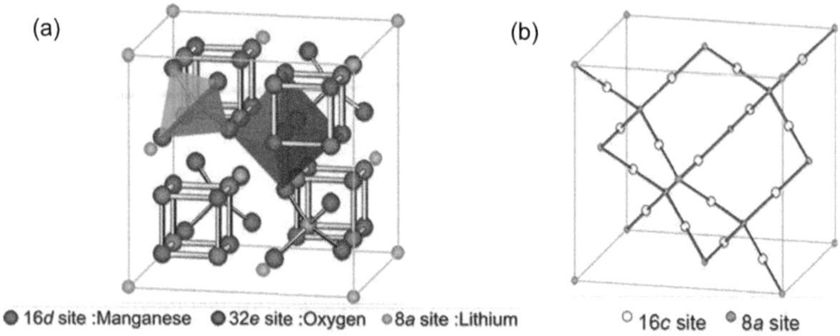

Figure 1.7 (a) Crystal structure of a $LiMn_2O_4$ spinel. (b) 8*a*–16*c* diffusion path of a Li ion in the spinel structure.

1.2.3 Olivine-Type Positive Electrodes

Positive electrode materials with an olivine-type structure are expressed as $LiMPO_4$ (M = Mn, Co, Ni, or Fe) and were first reported by Padhi et al. [30]. Among them, the most representative is $LiFePO_4$, which has the main advantage of using iron, which is inexpensive and abundant. The material also shows a comparatively high potential of approximately 3.5 V and has a theoretical capacity of 170 mAh/g. Moreover, it has been reported that it can actually obtain a capacity close to the theoretical value [31]. Therefore, from the standpoint of capacity, it is recognized as a material that can effectively surpass $LiCoO_2$. Because PO_4 forms a stable framework structure, $LiFePO_4$ has excellent thermal stability during charging and excellent cycling characteristics [31]. This material has recently attracted considerable attention due to these advantages, and active research is ongoing [32]. Figure 1.8 shows the crystal structure of $LiFePO_4$. The FeO_6 octahedra form a zigzag shape by corner sharing and forms parallel layers in the *bc* plane. These octahedra and PO_4 tetrahedra also have corner sharing, and in this way, a 3D network is formed. The Li atoms that exist at the octahedral sites form a chain in the direction of the *b* axis by edge sharing, and Li^+ diffuses along this chain. Charging and discharging proceed via a two-phase reaction of $LiFePO_4$ and $FePO_4$, so the potential is flat. In addition, as extremely highly efficient charging and discharging are possible, several charging and discharging mechanisms involving a two-phase region have been presented. For example, there have been many reports, such as a core-shell model in which the two-phase boundary moves within particles and models in which there is a solid solution region at high temperatures [33] or a solid solution region near room temperature [34] or models in which lithium diffuses in a region at the boundary of the two phases [35].

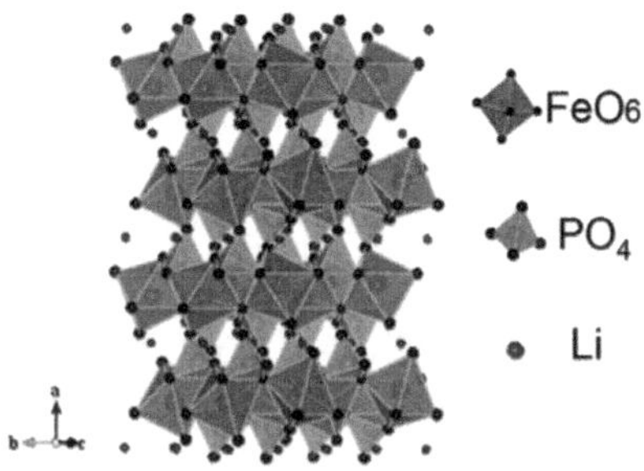

Figure 1.8 The crystal structure of a $LiFePO_4$ olivine structure.

Recently, there has also been research related to Li_2FeSiO_4, which uses Si instead of P [36, 37]. Similar to $LiFePO_4$, SiO_4 forms a stable framework, so the crystal structure is stable [36]. From the viewpoint of resources, Si is more abundant than P, and if both lithium ions can be extracted, it has the potential of showing a high capacity (330 mAh/g). Indeed, there are reports showing that it is possible to extract approximately two Li^+, resulting in a high capacity (330 mAh/g) [38].

1.2.4 Lithium-Rich Positive Electrode Materials

The first report that solid solutions of Li_2MnO_3 and $LiMO_2$ (M = Ni, Mn, or Co) show a high capacity was by Dahn, who in 2001, reported that Li_2MnO_3-$LiNiO_2$ solid solution materials have an irreversible plateau near 4.5 V and showed a high capacity (280 mAh/g) when charged as far as 4.8 V [39]. Later, Thackeray et al. reported that solid solutions of Li_2MnO_3 and $LiMn_{0.5}Ni_{0.5}O_2$ have a stable cycle with a high capacity [40], and they also reported the electrochemical characteristics of a solid solution of Li_2MnO_3 and

$LiNi_{1/3}Co_{1/3}Mn_{1/3}O_2$ [41] and elucidated the reaction mechanism using nuclear magnetic resonance spectroscopy [42]. However, at the present time, the mechanism for expressing a high capacity is still unclear. In addition, one major characteristic is that more lithium can be inserted and extracted reversibly and stably, compared to the usual layered materials, but the cause of this is unclear.

Armstrong et al. used a mass spectrometer during the initial charging period of a $Li[Ni_{0.2}Li_{0.2}Mn_{0.6}]O_2$-positive electrode material to show the dependency of the amount of O_2 gas generated on the charging voltage [43]. The charging capacity below 4.5 V was less than 100 mAh/g, but the charging capacity above 4.5 V reached more than 200 mAh/g. From the voltage dependency of the ion current due to the amount of oxygen gas generated from the positive electrode (measured for each 0.1 V starting from 4.3 V), it was found that oxygen evolution does not occur up to 4.5 V, starting at 4.6 V. Moreover, oxygen is generated in a range where the potential is flat. Furthermore, according to neutron beam diffraction, in the specimens charged to 4.4 V, only the lithium in the layer that is pure lithium is extracted from the initial structure, whereas the amount of lithium in the Mn-Li mixed layer does not change. One study examined in more detail the relationships involving the initial charging/discharging reversibility, including the process of oxygen extraction in this material system as well as the chemical composition and the average valence of the transition metal [44].

1.3 Negative Electrodes

The negative electrode–active material is a reductant. What is desired, as much as possible, is a material with a low potential, high reducing power, and a low electrochemical equivalent weight. Figure 1.9 shows the potential plotted as a function of the capacity of the negative electrode material. The capacity increases from graphite, to hard carbon, fine particle systems, nitrides, alloys, and finally metallic lithium, which is an extremely negative electrode material.

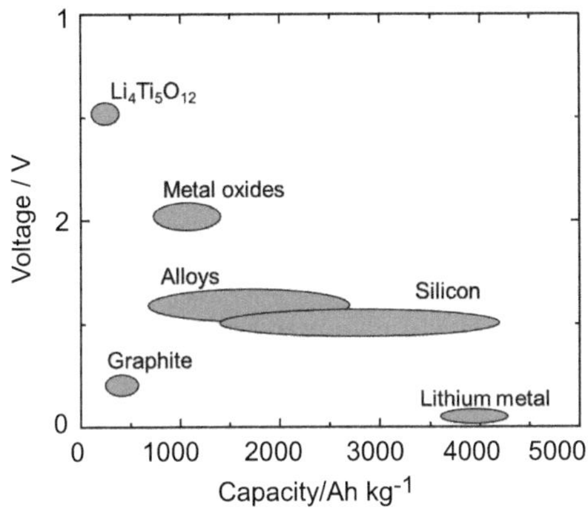

Figure 1.9 A roadmap of developing anode materials.

1.3.1 Graphite Negative Electrodes

For graphite carbon negative electrodes, the Li^+ insertion capacity is relatively large and the reversibility of the Li^+ insertion/extraction reaction is extremely high, so these are used widely as negative electrodes for LIBs. The following are the general characteristics of graphite materials:

- In contrast to metallic lithium, lithium deposition does not occur.
- The volumetric expansion ratio is small.
- The coulomb efficiency is high.
- Graphite materials exhibit high cycling characteristics even if the charging depth is 100%.
- The electrode potential is close to that of metallic lithium.
- There are few problems with cost or supply.
- There are no concerns related to people or the environment.

In this way, because they satisfy the important requirements for a battery on the market, they are used in almost all LIBs currently used. The capacity density per weight and mass with graphite is 372 mAh/g and 855 Ah/L, respectively, which are high values, even if they are not as high as those with metallic lithium, and also with regard to potential, a low value of 0.2 to 0.5 V (vs. Li/Li^+) is observed, which is roughly equivalent to that of a metallic lithium negative electrode.

In graphite, layers of hexagonal networks that have a strong structure due to covalent sp^2 bonding are stacked by van der Waals forces to form a polycrystalline material. Even with the insertion and extraction reactions of chemical species, the covalent sp^2 bonding network of the graphene plane is basically unchanged. This is a topochemical reaction system, in which Li^+ moves between the layers that are stacked in the *c* axis direction via van der Waals forces, so the graphite crystal structure has high stability during the insertion and extraction processes.

Accompanying the insertion of Li^+, the graphite negative electrode shows the same negative potential as metallic lithium, so there is a high reductive force, and many organic solvents would be decomposed reductively. Therefore, for thermodynamic reasons, organic solvents cannot be used as an electrolyte. When a solvent is used whose main component is propylene carbonate, which has been used in primary batteries, decomposition of the electrolyte would occur before the potential for the insertion of Li^+ is reached, so the insertion of Li^+ is electrochemically impossible. This problem was solved later using an electrolyte whose main component was ethylene carbonate [45]. The insertion of Li^+ becomes possible in this way with a different solvent due to the formation of a stable film (solid–electrolyte interface [SEI]) on the carbon surface, which deactivates the carbon surface [46]. This covering film displays Li^+ conduction but does not support electron conduction. Therefore, it suppresses further decomposition of the electrolyte, which makes Li^+ insertion into the carbon structure possible. Among the research into the interface between the graphite negative electrode and the electrolyte, there have been several reports on the composition and configuration of this type of film. For example, Besenhard et al. presented a solvent co-insertion model, in which the film is formed within the carbon material [47]. This model has the idea that solvated Li^+ is first inserted between the layers, which then experience reductive decomposition, and the decomposition products that remain between the layers function as the SEI. Furthermore, there have been attempts to improve the characteristics by modifying the carbon surface [48].

However, to form a good-quality SEI and maintain good cycling characteristics and battery safety, high-cost materials must be used for various parts of the battery. In addition, as water can have a deleterious effect during the SEI formation process, it is necessary to reduce the water content in the electrolyte to as little as 10 ppm, which also increases the cost of the electrolyte.

1.3.2 Negative Electrodes That Use a Conversion Reaction

Currently, electrodes in which the reaction progresses by Li^+ insertion and extraction are mainstream, but there is also research related to electrodes that perform Li^+ exchange via conversion reactions with Li^+. Most of the electrodes that work in this way show large-scale restructuring of the crystal lattice accompanying the Li^+ insertion/extraction reactions. In an early study, Thackeray et al. reported that with a LiCl-KCl-fused salt, Fe_2O_3 is reduced electrochemically to Fe and Li_2O during discharge and charging can be performed reversibly [49]. In addition, Denis et al. reported that in normal-temperature nonaqueous electrolytes, metal ions in certain types of vanadium oxides are reduced to the metallic form, which shows much higher reversible capacity than normal intercalation [50]. On the basis of this, Tarascon et al. presented a mechanism called a conversion reaction, in which a transition metal oxide and Li^+ are reacted reversibly [51, 52]. The equation for the reaction is as follows:

$$MO + 2Li^+ + 2e^- \rightleftarrows M + Li_2O \tag{1.3}$$

For this system, a "metal oxide" was first investigated, and as research into this reaction progressed, it was confirmed that similar reactions also occur when the oxygen ratio is higher. This reaction exists generally in all metal oxides. These types of reactions are complicated and vary according to the type of oxide, but the finally reduced products are basically mixtures of the metal M and Li_2O. Depending on the material, the reaction may not be complete under the condition on the right side because it is also possible that the deposited metal M reacts to form an alloy with lithium. Nevertheless, the basic characteristics common to the materials that perform the conversion reactions can be understood from the metal-oxide reaction equation Eq. 1.3.

A characteristic of using a conversion reaction at the electrode is that the capacity is much larger than that for the insertion-extraction type. On the other hand, because destruction and reconstruction of the crystal lattice accompany charging and discharging, in terms of reversibility, it is often inferior to the insertion type of reaction. In addition, metal oxides are compounds that have counteranions other than the metal ion, which is the main substance for charge transfer, so they do not generate a capacity as large as that generated by the alloy electrodes described later. However, the problem of volumetric expansion and contraction during charging and discharging, which occurs with alloy compounds, is reduced because the concentration of the metal itself is lower.

In addition to oxides, the reactions are known for nitrides such as Li_3N and phosphides [53] such as FeP_2 [54], in which Li^+ insertion and extraction accompany decrystallization of the crystal lattice. Considerable research has been performed on negative electrode materials, for example, $Li_{3-x}M_xN$ systems, in which electron conductivity is conferred by replacing a part of the Li in the Li_3N with a transition metal [55, 56].

1.3.3 Ti System Electrodes

Among the first transition series elements, Ti has the lowest M^+/M redox potential (−1.63 V vs. SHE), and the oxides that are its compounds also generate a low potential. Therefore, Ti oxides are often the subject of research for negative electrode materials. Among these, spinel oxides that have a stable crystal structure, as well as TiO_2, are often researched.

Spinel-type $Li_4Ti_5O_{12}$ is a typical lithium-containing titanium oxide. This material is known as a zero-strain material [57]. According to the X-ray diffraction measurements by Scharner et al., only 0.2% volumetric compression occurs when Li^+ is completely filled [58]. This compound displays an extremely flat potential profile at 1.55 V, but there is still controversy as to whether charging and discharging proceed with two coexisting phases or in a single-phase reaction. The capacity is approximately 150 mAh/g, and in terms of lithium conversion, this corresponds to 4 to 3.5 per formula weight being reduced. Considerable research has been carried out on this as a stable, high-speed material, because it is distortion free. $Li_4Ti_5O_{12}$ nanoparticles produced by the sol-gel method are processed into thin-film electrodes with a thickness of several micrometers, which introduces the ability to retain approximately 50% of the capacity at a high rate of 250 C [59].

1.3.4 Lithium Alloy Negative Electrodes

To avoid the generation of metallic lithium dendrites, a method of using lithium alloys has been considered. Dendrites are not believed to form if the surface is not lithium, so it is possible to avoid an internal short circuit.

A wide variety of metals, such as Sn, Pb, Al, Zn, Bi, Ga, In, and Si, have been assessed for alloys. In particular, Li-Al alloys have excellent properties. Therefore, their use as negative electrodes in secondary batteries has attracted considerable attention. The Li-Al alloy charging/discharging properties have an extremely flat charging/discharging plateau at 0.5 V (vs. Li/Li^+), and the capacity is approximately 800 mAh/g. The capacity is extremely small compared to that of the metallic lithium negative electrode (3860 mAh/g), but this is still very attractive considering that a graphite negative electrode is 372 mAh/g and that is currently the mainstream.

In 1976, there were reports of the use of Li-Si alloys as the negative electrodes in lithium secondary batteries that used molten salts as the electrolytes [60]. The theoretical capacity of the Li-Si alloys had the large value of 4000 mAh/g, but for this case, the operating temperature was over 400°C. Later, Wilson and Dahn reported that the capacity is also not achieved if Si powder is used at normal temperatures [61].

An alloy negative electrode has several disadvantages. First, similar to metallic lithium, the reactivity is high. Another problem is that large volumetric changes accompany the insertion and extraction of large amounts of lithium, so the electrode materials may be pulverized.

Therefore, with regard to practical applications of alloy negative electrodes, there has been considerable research and development into finely dispersing elements that react with lithium or form a matrix of elements that hold the reaction elements and do not react with lithium [62]. Recently, by mixing multiple elements, such as Sn, Co, and C, at the atomic level to make them amorphous, several attempts have been made to suppress the deformation of particles during charging and discharging and improve the cycle lifetime, and Sony has succeeded with practical application.

1.3.5 Metallic Lithium Negative Electrodes

Metallic lithium has an extremely high theoretical electrical capacity (3860 mAh/g) and a low redox potential (–3.04 V vs. SHE). In theory, it can almost be called the ultimate negative electrode material, but there are several problems with its practical application. The largest problem, as already mentioned, is that the reversibility of the charging/discharging reaction is extremely low. This is because the metallic lithium is not deposited homogeneously during charging and docs not dissolve reversibly during discharging. In addition, metallic lithium, which is highly reactive, has the danger of generating heat or igniting due to various causes, such as flames or water. In particular, for a secondary battery, the previously mentioned inhomogeneous deposition leads to metal interface structures (dendrites) with an extremely fine structure, such as needles, branches, or fine particles. These structures can penetrate the separator and reach the positive electrode, causing a short circuit. In the worst case, it can cause a large amount of heat or ignition due to sudden self-discharge.

The inhomogeneous deposition occurring during charging is the basic factor of the many problems with metallic lithium negative electrodes. Therefore, to make practical application possible, it is important to control the interface between lithium and electrolyte to improve the deposition conditions. In fact, the electrochemical deposition of a metal is always accompanied by the formation and growth of deposition nuclei, making it impossible to have the metallic lithium completely smooth down to the micro level. One practical solution is to control the macro and secondary structures that are formed as the deposited lithium grows, so that they give a condition that is safe with high reversibility, even if it is not homogeneous.

1.4 Electrolytes

Within a battery, the electrolyte is the most important component after the positive and negative electrodes. The basic properties of the battery, such as the operating voltage and the energy density, are determined theoretically by the chemistry of the positive and negative electrodes. However, for a practical battery, the choice of electrolyte often has a large effect on the characteristics of the battery. Lithium batteries have the issue that the high operating voltage leads to restrictions on the structure of the battery. On the other hand, a wide variation of electrolytes has been used to overcome such restrictions. To obtain excellent battery characteristics, the importance of performing appropriate electrolyte design has been recognized.

The following give some of the general properties that an electrolyte for a practical battery should have:

- High ion conductivity
- Wide potential window
- Thermal and chemical stability
- No chemical reaction with other materials within the battery, such as the active materials and the electrode collectors
- No or low toxicity
- Low cost

Currently, there are no electrolytes that satisfy all of these requirements for a lithium secondary battery. Rather, electrolytes are designed according to a battery's specifications and applications.

1.4.1 Organic Solvent Electrolytes

An organic solvent electrolyte is a system in which a lithium salt (supporting electrolyte) is dissolved in an organic solvent. A wide range of studies have been conducted since the introduction of lithium primary batteries, with various solvent/electrolyte combinations being proposed. They are used in most LIBs on the market today.

This section first discusses the solvent. Figure 1.10 presents representative organic solvents used in LIBs. When deciding on which solvent to use in a practical battery, properties such as the melting point, boiling point, dielectric constant, viscosity, number of donors, and number of acceptors need to be considered. The melting point and boiling points of the solvent are properties that directly affect the operating temperature range of the battery. For practical use, the operation must be safe both at high temperatures and at extremely low temperatures. In addition, the dielectric constant of the solvent is strongly related to properties such as the ionic dissociation and the association state of the electrolyte salt. Viscosity is an important property for determining ion mobility within the electrolyte, so it can have a large effect on Li^+ conductivity and the battery output. However, because it is possible to shorten the diffusion distance within the electrolyte by forming a battery in thin layers, it is rare for conductivity to determine the battery output when an organic solvent is used. The number of donors and the number of acceptors are parameters that show the nucleophilicity (basicity) and electrophilicity (acidity) of the solvent, respectively. These important indicators express the strength of the interaction with cations and anions (solvation). A large problem for organic solvents is the danger of ignition, so properties such as the flash point are also important from a safety point of view. Table 1.2 summarizes the properties of the main organic solvents.

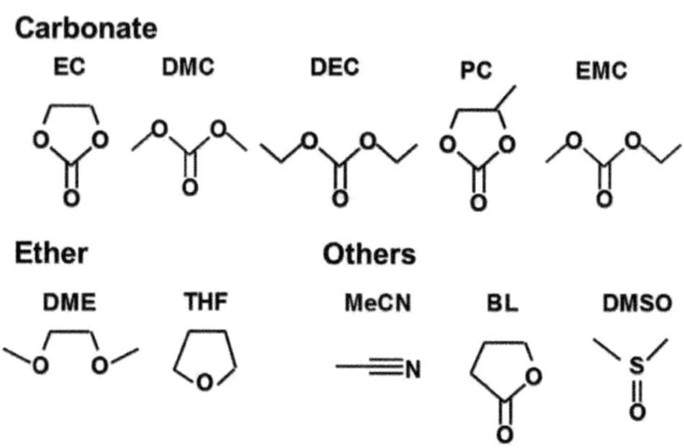

Figure 1.10 Molecular structures of organic solvents.

Table 1.2 Properties of various organic solvents

	Melting point T_m (°C)	Boiling point T_b (°C)	Density (g cm^{-3})	Dielectric constant (ε)
Dimethyl carbonate (DMC)	2.4	90	1.06	3.12
Diethyl carbonate (DEC)	−43.0	126	0.9752	2.82
g-butyrolactone (BL)	−43.3	204	1.1284	39.0
Propylene carbonate (PC)	−48.8	242	1.2047	66.14
Ethylene carbonate (EC)	36.4	248	1.3214	89.78

Next, lithium salt, which is the supporting electrolyte, is discussed. Table 1.3 lists the types and properties of lithium salts. Salts of univalent anions are desired because the ions easily dissociate in an organic solvent and have high ion mobility. Strong anions with small ionic radii have a strong electrostatic attraction to Li; hence, the solubility is low, or even if dissolved, the ions do not dissociate. Therefore, these generally cannot be used as electrolyte salts. $LiClO_4$ was used in lithium primary batteries, but a secondary battery is a strongly oxidizing environment during the charging process. Consequently, $LiClO_4$ has not been used in actual batteries due to safety considerations. From the viewpoint of solubility and chemical stability in organic solvents, among the salts based on Lewis acid fluorides, $LiBF_4$ and $LiPF_6$ are used in LIBs. In addition, various inorganic lithium salts based on Lewis acid fluorides as well as perfluoro organic sulfonates and imido salts have been proposed. The reason that so many candidates have been attempted is that there have been problems, such as conductivity and stability, even with $LiBF_4$ and $LIPF_6$ as battery electrolytes.

Table 1.3 Comparison of characteristics of typical lithium salts

	Average molecular weight	Melting point (°C)	Decomposition point (°C)	Conductivity in EC/DMC (mS cm^{-1})
$LiBF_4$	93.9	293	>100	4.9
$LiPF_6$	151.9	200	~80	10.7
$LiAsF_6$	195.9	340	>100	11.1
$LiClO_4$	106.4	236	>100	8.4
Li imide	286.9	234	>100	9.0
Li triflate	155.9	>300	>100	

1.4.2 Polymer Electrolytes

A battery that uses a polymer as the electrolyte is an all-solid system. Therefore, there is no danger of a liquid leak, and it is possible to make thin films and laminations. The polymer electrolytes that have been developed thus far can be roughly classified into dry polymer electrolytes and gel polymer electrolytes.

Research into dry polymer electrolytes began in 1975 with the report by Wright [63]. He showed that polyethylene oxide (PEO) forms a complex with an alkali metal in the solid state that displays ion conductivity at room temperature. The characteristics of the dry polymer structure are either straight-chain PEO or PEO included in such a structure, and polymers with polyester, polyamine, or polysulfide bonds have also been proposed.

Gel electrolytes are a general name for swelled bodies that have a 3D network structure as a result of the chemical bonds of a polymer or of molecular chain interactions, such as crystallization or molecular interweaving, that hold solvent molecules within the enclosures formed by the main backbone and side-chain molecules. These have a wide range of properties, ranging from being soft and having some fluidity to having some degree of mechanical strength. Ions move through the liquid within the gel, and the polymer itself serves both to maintain the mechanical strength of the gel overall and to maintain the liquid component. As a result, the conductivity of a gel polymer electrolyte is close to the conductivity of a liquid but higher than that of a dry polymer electrolyte.

1.4.3 Ionic Liquids

Materials that are made from cations and anions are generally solid at room temperature, as represented by NaCl. However, at the start of the 20th century, it was possible to obtain salts with amazingly low melting points, even those that were liquid at room temperature, that is, ionic liquids [64]. At the start of the 1990s, by giving cations and anions an appropriate molecular design, it was widely recognized for the first time that ionic liquids that are stable in both water and air could be obtained, and these attracted considerable attention as a third type of fluid, in addition to water and organic solvents. Ionic liquids have an extremely low vapor pressure and low combustibility. In addition, because they have high heat resistance, they are in the liquid state for a wide range of temperatures. They also have the characteristics of chemical stability and high ion conductivity. Another major feature of ionic liquids is that their properties can be controlled without limit according to the structural designs of the cations and anions and their combinations. With regard to applications to LIBs, there have been many reports of the use of ionic liquids composed of perfluoro anions [65, 66]. However, the potential window is small on the reduction side. Matsumoto et al. succeeded in greatly improving the reduction resistance over previous ionic liquids by synthesizing an ionic liquid that uses an aliphatic quaternary ammonium salt [67]. Ishikawa et al. also reported reversible electrochemical Li^+ insertion/extraction reactions for graphite electrodes using bis(fluoro)sulfonylimide, a new anion, for the ionic liquid [66].

1.4.4 Inorganic Solid Electrolytes

When an inorganic solid electrolyte is used, the structure of the battery system can be simplified so that the total energy density is increased. These materials also have high stability against electrochemical and thermal decomposition. Therefore, they have recently been assessed as inorganic solid electrolytes for next-generation batteries for electric vehicles.

The inorganic solid electrolytes that are currently being researched can be divided mainly into two types, oxide solid electrolytes and sulfide solid electrolytes. Oxide solid electrolytes do not react with oxygen or water, so they make few demands on the environment during production and application. However, because oxides are hard and brittle, there are problems in the solid-solid junction with the positive electrode or the negative electrode, which are also solids, and in fact, it is difficult to create a cell. On the other hand, in a sulfide solid electrolyte, in which a sulfide is used, it reacts with water in the air to generate poisonous hydrogen sulfide. Therefore, strict management of the environment is required during production and application. Despite this, a sulfide is softer than an oxide, so contact with the electrodes is easier, and the ion conductivity is higher than that with oxide solid electrolytes, and sulfides have recently been attracting attention.

For application in a bulk solid battery, an electrolyte that shows a conductivity of at least 10^{-3} S cm^{-1} is preferred. Among the oxide crystals available, systems that show high conductivity (10^{-4} to 10^{-3} S cm^{-1}) at room temperature, such as the perovskite-type $La_{0.51}Li_{0.34}TiO_{2.94}$, the NASICON-type $Li_{1.3}Al_{0.3}Ti_{1.7}(PO_4)_3$, and the garnet-type $Li_7La_3Zr_2O_{12}$, are known. The development of sulfide electrolytes is progressing, the focus being on glasses. In two-component systems, such as Li_2S-SiS_2 systems, and three-component systems that also add in LiI or Li_3PO_4, a materials search has been performed, and glasses that show a conductivity of at least 10^{-3} S cm^{-1} at room temperature have been obtained.

1.5 Reactions of Lithium-Ion Batteries

The reaction front in today's LIBs is a Li^+ transfer reaction at an electrode–electrolyte heterointerface. Compared to the studies related to material structures, research related to real charge transfer reactions at the interfaces is lagging. Elucidation of the fundamental mechanism of Li^+ transfer reactions at the interfaces will certainly be a key to improving the performance of LIBs and the design of the next-generation secondary batteries.

The redox reactions that occur at the electrode surface are inhomogeneous reactions. Therefore, the charge transfer reaction, which is the chemical reaction that accompanies the transfer of charged species, may become the rate-determining process. However, the amounts of reacting materials that are supplied to the reaction front, that is, the flux of reactants from the bulk phase to the reaction front, can be the rate-determining process. In the latter case, the mass transport by diffusion is rate determining, and because the charge transfer reaction is relatively fast, it is believed that the charge transfer reaction will progress to reach a state that is visible at equilibrium. On the other hand, when the charge transfer reaction rather than mass transport is rate determining, the kinetics of the charge transfer reactions become important (Fig. 1.11).

A reaction, such as that shown in Eq. 1.4, progresses at the electrode surface:

$$\mathrm{Ox} + ne^- \rightleftarrows \mathrm{Red} \tag{1.4}$$

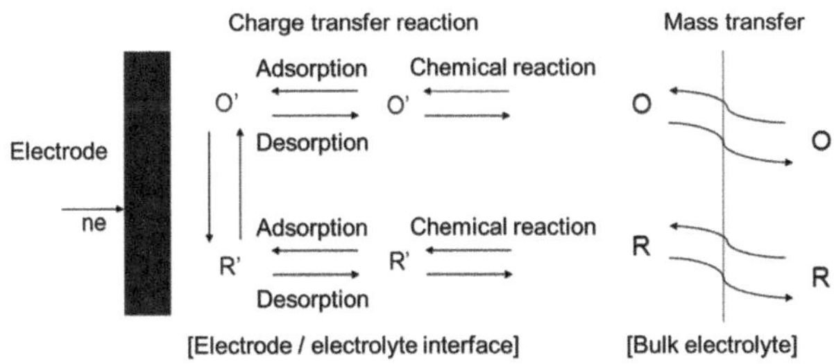

Figure 1.11 Schematic figures of a charge transfer reaction and mass transfer.

Ox and Red indicate the oxidant and the reductant that participate in the electrode reaction, e^- indicates an electron, and n is the number of electrons that participate in the electrode reaction. In Eq. 1.5, k_f and k_b are the rate constants for the forward and backward reactions, respectively. In this case, if C_{Ox} and C_{Red} are the concentrations of the oxidant and reductant, respectively, then the reaction rates can be expressed as:

$$v_f = k_f C_{Ox} \tag{1.5}$$

$$v_b = k_b C_{Red} \tag{1.6}$$

If the activation energies are ΔG_f^* and ΔG_b^* and the frequency factors are A_f and A_b, then the rate constants of the respective electrode reactions, according to transition state theory, can be expressed as:

$$k_f = A_f \exp(-\Delta G_f^*/RT) \tag{1.7}$$

$$k_b = A_b \exp(-\Delta G_b^*/RT) \tag{1.8}$$

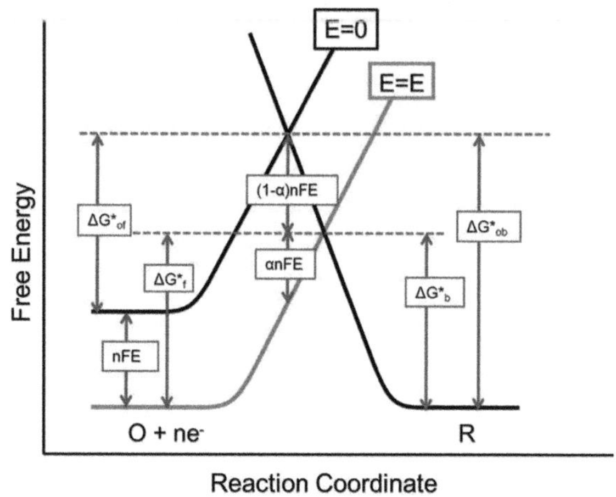

Figure 1.12 Potential energy curves of charge transfer.

The potential energy can be converted to a potential, and if the electrode potential is displaced by E in the positive direction from the state, in which the electrode potential is zero ($E = 0$) relative to an arbitrary standard for the potential, then the potential energy is displaced by nFE in the negative direction. On the other hand, if the electrode potential is changed by E in the negative direction, then the potential energy is displaced in nFE in the positive direction. In Fig. 1.12, the solid line and the dotted line are the potential energy curves for cases when the electrode potential is 0 or is displaced by E, respectively. If the activation energies for the forward and backward directions when the electrode potential is 0 are ΔG_{0f}^* and ΔG_{0b}^*, respectively, and the activation energies for the forward and backward reactions when the potential is E are $\Delta G_f^*$and ΔG_b^*, respectively, then the following equations can be obtained:

$$\Delta G_f^* = \Delta G_{0f}^* + anFE \tag{1.9}$$

$$\Delta G_b^* = \Delta G_{0b}^* - (1 - a)nFE \tag{1.10}$$

Here, a is called the transfer coefficient, which is a parameter that shows the fractional contribution of the electrode potential to the activation energy, and it is in the range $0 < a < 1$, but is generally near 0.5.

If Eqs. 1.9 and 1.10 are substituted into Eqs. 1.7 and 1.8, respectively, the following equations are derived:

$$k_f = A_f \exp(-\Delta G_{0f}^{*}/RT)\exp(-\alpha nfE) \quad (1.11)$$

$$k_b = A_b \exp(-\Delta G_{0b}^{*}/RT)\exp|(1-\alpha)nfE| \quad (1.12)$$

Here, $f = F/RT$. $A_f\exp(-\Delta G_{0f}^{*}/RT)$ and $A_b\exp(-\Delta G_{0b}^{*}/RT)$ in Eqs. 1.11 and 1.12 are not related to the potential and are equal to the rate constants when $E = 0$. If these are set to k_f^0 and k_b^0, respectively, the following can be obtained:

$$k_f = k_f^0 \exp(-\alpha nfE) \quad (1.13)$$

$$k_b = k_b^0 \exp|(1-\alpha)nfE| \quad (1.14)$$

Next, if the activity of the oxidant and the oxidant are denoted by a_{Ox} and a_{Red} then the respective reaction rates, v_f and v_b, are:

$$v_f = k_f a_{Ox} \quad (1.15)$$

$$v_b = k_b a_{Red} \quad (1.16)$$

For the equilibrium potential condition ($E = E^0$), from the Nernst equation,

$$E^{eq} = E^{0'} + \frac{RT}{nF}\ln\left(\frac{C_{Ox}}{C_{Red}}\right) = E^0 + \frac{RT}{nF}\ln\left(\frac{a_{Ox}}{a_{Red}}\right), \quad (1.17)$$

where $E^{0'}$ is the formal potential of an electrode. In addition, if the active amounts of oxidant and reductant are the same under the equilibrium conditions then the reaction rates for oxidation and reduction are also the same; hence,

$$k_f a_{Ox} = k_b a_{Red} \quad (1.18)$$

and $k_f = k_b$. In this case, if the rate constant is set to k^0,

$$k^0 = k_f^0 \exp\left(-\alpha nfE^0\right) = k_b^0 \exp|(1-\alpha)nfE^0| \quad (1.19)$$

Therefore, k^0 is equal to the rate constants for both the forward reaction and the backward reaction under standard conditions. If the above are substituted into Eqs. 1.13 and 1.14, then the following equations can be derived:

$$k_f = k^0 \exp|(-\alpha nf(E - E^0)| \quad (1.20)$$

$$k_b = k^0 \exp|(1-\alpha)nf(E-E^0)| \tag{1.21}$$

Each Faraday current value is expressed as the multiplication of the rate constant, the active amount of the oxidant or reductant, the number of electrons involved in the reaction, and Faraday's constant. The total current is the difference between the Faraday currents for the forward and backward reactions. Therefore,

$$i = nFk^0 \left\{a_{\text{Ox}} \exp|-\alpha nf(E-E^0)| - a_{\text{Red}} \exp|(1-\alpha)nf(E-E^0)|\right\} \tag{1.22}$$

This is called the Butler–Volmer equation, and it is extremely important in electrode reaction theory because it is the basic equation that shows the contribution of the electrostatic energy to the charge transfer rate. This equation applies when only the charge transfer process determines the current density through the system.

At the equilibrium potential, the potential is $E = E^{\text{eq}}$, and in that case the total current that flows through the system is 0. This is expressed using Eq. 1.22 as follows:

$$nFk^0 a_{\text{Ox}} \exp|-\alpha nf(E^{\text{eq}}-E^0)| = nFk^0 a_{\text{Red}} \exp|(1-\alpha)nf(E^{eq}-E^0)|. \tag{1.23}$$

It is not that the current is 0 but rather that the Faraday currents for the forward and backward reactions have the same magnitude and are therefore balanced. This current value is called the exchange current, and in this case, the current density is expressed as the exchange current density, i_0:

$$i_0 = nFk^0 a_{\text{Ox}} \exp|-\alpha nf(E^{\text{eq}}-E^0)| = nFk^0 a_{\text{Red}} \exp|(1-\alpha)nf(E^{eq}-E^0)| \tag{1.24}$$

If Eq. 1.23 is substituted for ($E^{\text{eq}} - E^0$) in Eq. 1.17 and simplified, the following can be obtained:

$$i_0 = nFk^0 (a_{\text{Ox}})^{1-\alpha} (a_{\text{Red}})^{\alpha} \tag{1.25}$$

Butler [68] defined the concept of the exchange current in 1936, and it is extremely useful for comparing the reaction rates at the equilibrium potentials. Therefore, it is often used instead of k^0 as the kinetics parameter in the electrode reaction.

At the electrode–electrolyte interface, the potential determined by the difference between the chemical potentials of the reacting species (the equilibrium electrode potentials) applies and charge separation occurs. Owing to this charge separation, a very thin region, called the electrical double layer, which has a potential gradient, exists on the electrolyte side of the electrode–electrolyte interface. Figure 1.13 presents the Grahame electrical double-layer model. A layer of specific absorption ions is formed closest to the electrode, which is called the inner Helmholtz layer. Outside the inner Helmholtz layer, a layer of ions that have an opposite charge from the specific absorption ions (the outer Helmholtz layer) is formed. The actual battery reaction proceeds in this electrical double layer, and phenomena such as the absorption and exchange of electrons, the reactions and transfer of ions, and the change in the electrode structure occur at the interface.

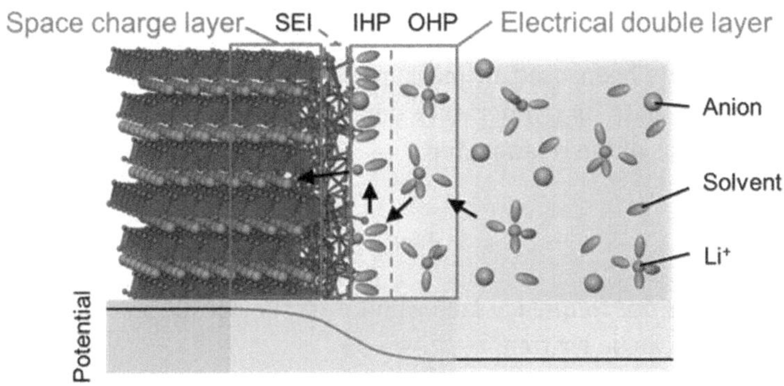

Figure 1.13 Schematic view of an electrode–electrolyte interface. SEI, IHP, and OHP mean the solid–electrolyte interface, the inner Helmholtz plane, and the outer Helmholtz plane, respectively.

Here, the charge transfer reaction at the electrode–electrolyte interface is considered. If the charge transfer reaction at the interface is assumed to follow the Butler–Volmer equation, then the charge transfer resistance (R_{ct}) can be expressed as

$$R_{ct} = \frac{RT}{nFAi_0} \tag{1.26}$$

where i_0 is the exchange current density, R is the gas constant, T is the absolute temperature, n is the number of electrons in the reaction, and F is Faraday's constant. If the lithium-ion active amounts within the electrolyte and the electrode are ${a_{\mathrm{Li}}}^{\mathrm{sol}}$ and ${a_{\mathrm{Li}}}^{\mathrm{ele}}$, respectively, then the exchange current density can be expressed as [69]:

$$i_0 = k^0 \left(a_{\mathrm{Li}}^{\mathrm{sol}}\right)^{1-\alpha} \left(a_{\mathrm{Li}}^{\mathrm{ele}}\right)^{\alpha} \tag{1.27}$$

The charge transfer reaction that proceeds in a LIB, which is different from a reaction due to electron transfer, can be interpreted as an interphase transfer reaction, in which Li^+ is transferred across the electrode–electrolyte interface. The above equation is derived from an electron transfer reaction. Therefore, when discussing a Li^+ transfer reaction, it is necessary to verify in detail all the factors within the equation.

One method of measuring the charge transfer resistance is the electrochemical impedance spectroscopy (EIS) method. The EIS method has been used to measure the charge transfer resistance and investigate its temperature dependence [70, 71]. In particular, on the basis of the relationship

$$i_0 = A \exp\left(-\frac{E_{\mathrm{a}}}{RT}\right) \tag{1.28}$$

where A is the frequency factor and E_{a} is the activation energy, the following can be written:

$$\frac{1}{R_{\mathrm{ct}}} \propto i_0 = A \exp\left(-\frac{E_{\mathrm{a}}}{RT}\right) \tag{1.29}$$

Measurements of the activation energy of the charge transfer reaction provide important information for extracting the factors that control the reaction occurring in the system. However, when the activation energy of a charge transfer reaction is measured using an electrode/liquid electrolyte system, it is still unclear as to at what state in the charge transfer reaction process the value should be measured. Bruce and Saidi presented an “ad atom” model [72, 73] (Fig. 1.14), which modifies the intercalation reaction system, and the following reactions proceed in the process of intercalation from the electrolyte into the electrode:

1. Solvated lithium ions diffuse through the electrolyte liquid to the electrode surface.
2. Lithium ions are absorbed at the electrode surface.
3. The solvated ions that are solvated in lithium are extracted.
4. The charge transfer process and surface conduction occur.
5. Lithium ions are inserted into the electrode.
6. Lithium ions diffuse through the electrode.

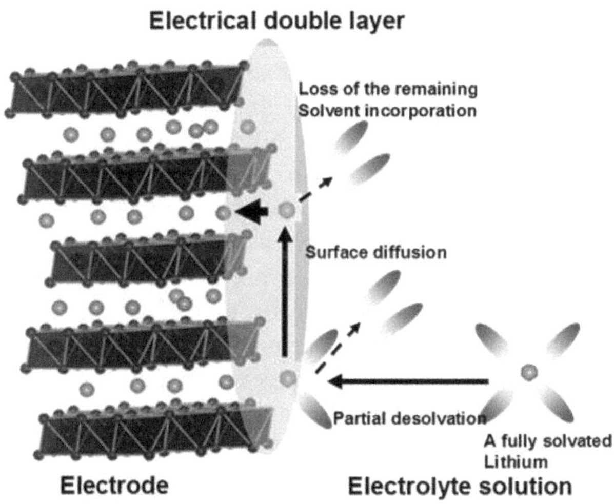

Figure 1.14 Schematic model of a charge transfer reaction based on the adion model.

Among these processes, the desolvation process (3) and the process whereby the desolvated Li^+ (adions) are inserted into the electrode (5) appear to have large activation barriers.

If the solvent is the same, then for any solid material, such as a carbon thin film [74], carbon sheets [75], or a $LiMn_2O_4$ electrode [76], the activation energy is similar. This strongly suggests that the main factor that controls the activation energy in the charge transfer reaction process is the desolvation process.

Obtaining information on the electrode–electrolyte interface is extremely important for future storage battery development. However, using theoretical calculations to analyze the phenomena at the electrode–electrolyte interface was previously difficult due to the complicated configuration, in which various species, such as ions, atoms, and solvent molecules, interact. The electrical double layer that constitutes the electrode–electrolyte interface exists as a very thin region of several nanometers, and methods for observing such an interface structure at the nanometer level have not been established. For these reasons, a new method that can observe the electrode–electrolyte interface directly is needed.

1.6 Outlook

In this chapter, we outlined the LIB and explained its components and reaction mechanism. In the future, further improvement of energy density is required for LIBs, and development of new secondary batteries is underway. In the next chapter, we will explain the development situation of post-LIBs that can be expected to have high energy density. Further, in the reaction of the secondary battery, continuous multiscale reaction proceeds in a nonequilibrium state, and this quality directly leads to performance. Therefore, it is necessary to understand internal phenomena to dramatically improve battery performance. However, since the secondary battery has a sealed structure, its observation is not easy. In Chapter 3, focusing on analysis of battery reactions, we will explain multiscale reaction phenomena. In Chapter 4 and later, we will explain in detail the ion beam analysis, which can help us directly observe the composition change of lithium ions.

References

1. Ikeda, H., Saito, T. and Tamura, H. (1975). *Manganese dioxide as cathodes for lithium batteries, Manganese Dioxide Symposium*, Cleveland, **1**, p. 384.
2. Whittingham, M. S. (1976). Electrical energy-storage and intercalation chemistry, *Science*, **192**, pp. 1126–1127.
3. Nagaura, T. and Tazawa, K. (1990). Lithium ion rechargeable battery, *Progress in Batteries and Solar Cells*, **9**, pp. 209–217.
4. Tarascon, J.-M. and Armand, M. (2001). Issues and challenges facing rechargeable lithium batteries, *Nature*, **414**, pp. 359–367.
5. Ohzuku, T. and Brodd, R. J. (2007). An overview of positive-electrode materials for advanced lithium-ion batteries, *J. Power Sources*, **174**, pp. 449–456.

6. Murphy, D. W. and Trumbore, F. A. (1976). The chemistry of TiS_3 and $NbSe_3$ cathodes, *J. Electrochem. Soc.*, **123**, pp. 960–964.

7. Mizushima, K., Jones, P. C., Wiseman, P. J. and Goodenough, J. B. (1980). Li_xCoO_2 (0<x<-1): a new cathode material for batteries of high energy density, *Mater. Res. Bull.*, **15**, pp. 783–789.

8. Thomas, M. G. S. R., David, W. I. F., Goodenough, J. B. and Groves, P. (1985). Synthesis and structural characterization of the normal spinel $LiNi_2O_4$, *Mater. Res. Bull.*, **20**, pp. 1137–1146.

9. Thackeray, M. M., Johnson, P. J., Depicciotto, L. A., Bruce, P. G. and Goodenough, J. B. (1984). Electrochemical extraction of lithium from $LiMn_2O_4$, *Mater. Res. Bull.*, **19**, pp. 179–187.

10. Levi, M. D., Salitra, G., Markovsky, B., Teller, H., Aurbach, D., Heider, U. and Heider, L. (1999). Solid-state electrochemical kinetics of Li-ion intercalation into $Li_{1-x}CoO_2$: simultaneous application of electroanalytical techniques SSCV, PITT, and EIS, *J. Electrochem. Soc.*, **146**, pp. 1279–1289.

11. Kano, S. and Sato, M. (1995). Structure and lithium insertion characteristics of $LiCrMnO_4$, *Solid State Ionics*, **79**, pp. 215–219.

12. Koksbang, R., Barker, J., Shi, H. and Saïdi, M. Y. (1996). Cathode materials for lithium rocking chair batteries, *Solid State Ionics*, **84**, pp. 1–21.

13. Rabou, L. P. L. M. and Roskam, A. (1995). Cycle-life improvement of $LiLiCoO_2$ batteries, *J. Power Sources*, **54**, pp. 316–318.

14. Takada, K., Aotani, N., Iwamoto, K. and Kondo, S. (1995). Electrochemical behavior of Li_xMO_2 (M = Co, Ni) in all solid state cells using a glass electrolyte, *Solid State Ionics*, **79**, pp. 284–287.

15. Reimers, J. N. and Dahn, J. R. (1992). Electrochemical and in situ X-ray-diffraction studies of lithium intercalation in Li_xCoO_2, *J. Electrochem. Soc.*, **139**, pp. 2091–2097.

16. Delmas, C. (1989). Alkali metal intercalation in layered oxides, *Mater. Sci. Eng., B*, **3**, pp. 97–101.

17. Liu, G. Q., Kuo, H. T., Liu, R. S., Shen, C. H., Shy, D. S., Xing, X. K. and Chen, J. M. (2010). Study of electrochemical properties of coating ZrO_2 on $LiCoO_2$, *J. Alloys Compd.*, **496**, pp. 512–516.

18. Li, C., Zhang, H. P., Fu, L. J., Liu, H., Wu, Y. P., Rahm, E., Holze, R. and Wu, H. Q. (2006). Cathode materials modified by surface coating for lithium ion batteries, *Electrochim. Acta*, **51**, pp. 3872–3883.

19. Ohzuku, T. and Makimura, Y. (2001). Layered lithium insertion material of $LiNi_{1/2}Mn_{1/2}O_2$: a possible alternative to $LiCoO_2$ for advanced lithium-ion batteries, *Chem. Lett.*, **30**, pp. 744–745.

20. Yabuuchi, N. and Ohzuku, T. (2003). Novel lithium insertion material of $LiCo_{1/3}Ni_{1/3}Mn_{1/3}O_2$ for advanced lithium-ion batteries, *J. Power Sources*, **119**, pp. 171–174.

21. Ohzuku, T. and Makimura, Y. (2001). Layered lithium insertion material of $LiCo_{1/3}Ni_{1/3}Mn_{1/3}O_2$ for lithium-ion batteries, *Chem. Lett.*, **30**, pp. 642–643.

22. Myung, S.-T., Amine, K. and Sun, Y.-K. (2010). Surface modification of cathode materials from nano- to microscale for rechargeable lithium-ion batteries, *J. Mater. Chem.*, **20**, pp. 7074–7095.

23. Yoon, W. S., Balasubramanian, M., Yang, X. Q., Fu, Z. G., Fischer, D. A. and McBreen, J. (2004). Soft X-ray absorption spectroscopic study of a $LiNi_{0.5}Mn_{0.5}O_2$ cathode during charge, *J. Electrochem. Soc.*, **151**, pp. A246–A251.

24. Berg, H., Kelder, E. M. and Thomas, J. O. (1999). Neutron diffraction study of stoichiometric spinel $Li_{1+x}Mn_{2-x}O_4$ showing octahedral 16c-site Li-occupation, *J. Mater. Chem.*, **9**, pp. 427–429.

25. Gummow, R. J., Dekock, A. and Thackeray, M. M. (1994). Improved capacity retention in rechargeable 4 V lithium lithium manganese oxide (spinel) cells, *Solid State Ionics*, **69**, pp. 59–67.

26. Xia, Y. Y., Zhou, Y. H. and Yoshio, M. (1997). Capacity fading on cycling of 4 V $Li/LiMn_2O_4$ cells, *J. Electrochem. Soc.*, **144**, pp. 2593–2600.

27. Wakihara, M., Li, G. H., Ikuta, H. and Uchida, T. (1996). Chemical diffusion coefficients of lithium in $LiM_yMn_{2-y}O_4$ (M = Co and Cr), *Solid State Ionics*, **86–88**, pp. 907–909.

28. Kawai, H., Nagata, M., Kageyama, H., Tukamoto, H. and West, A. R. (1999). 5 V lithium cathodes based on spinel solid solutions $Li_2Co_{1+X}Mn_{3-X}O_8$: $-1 \leq X \leq 1$, *Electrochim. Acta*, **45**, pp. 315–327.

29. Alcantara, R., Jaraba, M., Lavela, P. and Tirado, J. L. (2002). Optimizing preparation conditions for 5 V electrode performance, and structural changes in $Li_{1-x}Ni_{0.5}Mn_{1.5}O_4$ spinel, *Electrochim. Acta*, **47**, pp. 1829–1835.

30. Padhi, A. K., Nanjundaswamy, K. S. and Goodenough, J. B. (1997). Phospho-olivines as positive-electrode materials for rechargeable lithium batteries, *J. Electrochem. Soc.*, **144**, pp. 1188–1194.

31. Yamada, A., Chung, S. C. and Hinokuma, K. (2001). Optimized $LiFePO_4$ for lithium battery cathodes, *J. Electrochem. Soc.*, **148**, pp. A224–A229.

32. Malik, R., Zhou, F. and Ceder, G. (2011). Kinetics of non-equilibrium lithium incorporation in $LiFePO_4$, *Nat. Mater.*, **10**, pp. 587–590.

33. Delacourt, C., Poizot, P., Tarascon, J. M. and Masquelier, C. (2005). The existence of a temperature-driven solid solution in Li_xFePO_4 for 0 < x < 1, *Nat. Mater.*, **4**, pp. 254–260.

34. Yamada, A., Koizumi, H., Nishimura, S. I., Sonoyama, N., Kanno, R., Yonemura, M., Nakamura, T. and Kobayashi, Y. (2006). Room-temperature miscibility gap in Li_xFePO_4, *Nat. Mater.*, **5**, pp. 357–360.

35. Chen, G. Y., Song, X. Y. and Richardson, T. J. (2007). Metastable solid-solution phases in the $LiFePO_4/FePO_4$ system, *J. Electrochem. Soc.*, **154**, pp. A627–A632.

36. Nyten, A., Abouimrane, A., Armand, M., Gustafsson, T. and Thomas, J. O. (2005). Electrochemical performance of Li_2FeSiO_4 as a new Li-battery cathode material, *Electrochem. Commun.*, **7**, pp. 156–160.

37. Nishimura, S. I., Hayase, S., Kanno, R., Yashima, M., Nakayama, N. and Yamada, A. (2008). Structure of Li_2FeSiO_4, *J. Am. Chem. Soc.*, **130**, pp. 13212–13213.

38. Masese, T., Tassel, C., Orikasa, Y., Koyama, Y., Arai, H., Hayashi, N., Kim, J., Mori, T., Yamamoto, K., Kobayashi, Y., Kageyama, H., Ogumi, Z. and Uchimoto, Y. (2015). Crystal structural changes and charge compensation mechanism during two lithium extraction/insertion between Li_2FeSiO_4 and $FeSiO_4$, *J. Phys. Chem. C*, **119**, pp. 10206–10211.

39. Lu, Z., MacNeil, D. D. and Dahn, J. R. (2001). Layered cathode materials $Li[Ni_xLi_{(1/3-2x/3)}Mn_{(2/3-x/3)}]O_2$ for lithium-ion batteries, *Electrochem. Solid-State Lett.*, **4**, pp. A191–A194.

40. Johnson, C. S., Kim, J. S., Lefief, C., Li, N., Vaughey, J. T. and Thackeray, M. M. (2004). The significance of the Li_2MnO_3 component in 'composite' $_xLi_{(2)}MnO_{(3)}$ center dot $_{(1-x)}LiMn_{0.5}Ni_{0.5}O_2$ electrodes, *Electrochem. Commun.*, **6**, pp. 1085–1091.

41. Thackeray, M. M., Johnson, C. S., Vaughey, J. T., Li, N. and Hackney, S. A. (2005). Advances in manganese-oxide 'composite' electrodes for lithium-ion batteries, *J. Mater. Chem.*, **15**, pp. 2257–2267.

42. Thackeray, M. M., Kang, S. H., Johnson, C. S., Vaughey, J. T. and Hackney, S. A. (2006). Comments on the structural complexity of lithium-rich $Li_{1+x}M_{1-x}O_2$ electrodes (M = Mn, Ni, Co) for lithium batteries, *Electrochem. Commun.*, **8**, pp. 1531–1538.

43. Armstrong, A. R., Holzapfel, M., Novák, P., Johnson, C. S., Kang, S.-H., Thackeray, M. M. and Bruce, P. G. (2006). Demonstrating oxygen loss and associated structural reorganization in the lithium battery cathode $Li[Ni_{0.2}Li_{0.2}Mn_{0.6}]O_2$, *J. Am. Chem. Soc.*, **128**, pp. 8694–8698.

44. Arunkumar, T. A., Alvarez, E. and Manthiram, A. (2007). Structural, chemical, and electrochemical characterization of layered $Li[Li_{0.17}Mn_{0.33}Co_{0.5-y}Ni_y]O_2$ cathodes, *J. Electrochem. Soc.*, **154**, pp. A770–A775.

45. Fong, R., von Sacken, U. and Dahn, J. R. (1990). Studies of lithium intercalation into carbons using nonaqueous electrochemical cells, *J. Electrochem. Soc.*, **137**, pp. 2009–2013.

46. Yazami, R. and Guerard, D. (1993). Some aspects on the preparation, structure and physical and electrochemical properties of Li_xC_6, *J. Power Sources*, **43**, pp. 39–46.

47. Besenhard, J. O., Winter, M., Yang, J. and Biberacher, W. (1995). Filming mechanism of lithium-carbon anodes in organic and inorganic electrolytes, *J. Power Sources*, **54**, pp. 228–231.

48. Peled, E., Menachem, C., BarTow, D. and Melman, A. (1996). Improved graphite anode for lithium-ion batteries - chemically bonded solid electrolyte interface and nanochannel formation, *J. Electrochem. Soc.*, **143**, pp. L4–L7.

49. Thackeray, M. M. and Coetzer, J. (1981). A preliminary investigation of the electrochemical performance of α-Fe_2O_3 and Fe_3O_4 cathodes in high-temperature cells, *Mater. Res. Bull.*, **16**, pp. 591–597.

50. Denis, S., Baudrin, E., Orsini, F., Ouvrard, G., Touboul, M. and Tarascon, J. M. (1999). Synthesis and electrochemical properties of numerous classes of vanadates, *J. Power Sources*, **81–82**, pp. 79–84.

51. Poizot, P., Laruelle, S., Grugeon, S., Dupont, L. and Tarascon, J. M. (2001). Searching for new anode materials for the Li-ion technology: time to deviate from the usual path, *J. Power Sources*, **97–98**, pp. 235–239.

52. Poizot, P., Laruelle, S., Grugeon, S., Dupont, L. and Tarascon, J. M. (2000). Nano-sized transition-metaloxides as negative-electrode materials for lithium-ion batteries, *Nature*, **407**, pp. 496–499.

53. Lapp, T., Skaarup, S. and Hooper, A. (1983). Ionic-conductivity of pure and doped Li_3N, *Solid State Ionics*, **11**, pp. 97–103.

54. Silva, D. C. C., Crosnier, O., Ouvrard, G., Greedan, J., Safa-Sefat, A. and Nazar, L. F. (2003). Reversible lithium uptake by FeP_2, *Electrochem. Solid State Lett.*, **6**, pp. A162–A165.

55. Nishijima, M., Tadokoro, N., Takeda, Y., Imanishi, N. and Yamamoto, O. (1994). Li deintercalation-intercalation reaction and structural-change in lithium transition-metal nitride, Li_7MnN_4, *J. Electrochem. Soc.*, **141**, pp. 2966–2971.

56. Nishijima, M., Takeda, Y., Imanishi, N., Yamamoto, O. and Takano, M. (1994). Li deintercalation and structural-change in the lithium transition-metal nitride Li_3FeN_2, *J. Solid State Chem.*, **113**, pp. 205–210.

57. Ohzuku, T., Ueda, A. and Yamamoto, N. (1995). Zero-strain insertion material of $Li[Li_{1/3}Ti_{5/3}]O_4$ for rechargeable lithium cells, *J. Electrochem. Soc.*, **142**, pp. 1431–1435.

58. Scharner, S., Weppner, W. and Schmid-Beurmann, P. (1999). Evidence of two-phase formation upon lithium insertion into the $Li_{1.33}Ti_{1.67}O_4$ spinel, *J. Electrochem. Soc.*, **146**, pp. 857–861.

59. Kavan, L. and Gratzel, M. (2002). Facile synthesis of nanocrystalline $Li_4Ti_5O_{12}$ (spinel) exhibiting fast Li insertion, *Electrochem. Solid State Lett.*, **5**, pp. A39–A42.

60. Lai, S. C. (1976). Solid lithium-silicon electrode, *J. Electrochem. Soc.*, **123**, pp. 1196–1197.

61. Wilson, A. M. and Dahn, J. R. (1995). Lithium insertion in carbons containing nanodispersed silicon, *J. Electrochem. Soc.*, **142**, pp. 326–332.

62. Idota, Y., Kubota, T., Matsufuji, A., Maekawa, Y. and Miyasaka, T. (1997). Tin-based amorphous oxide: a high-capacity lithium-ion-storage material, *Science*, **276**, pp. 1395–1397.

63. Wright, P. V. (1975). Electrical conductivity in ionic complexes of poly(ethylene oxide), *Br. Polym. J*, **7**, p. 319.

64. Walden, P. (1914). Molecular weights and electrical conductivity of several fused salts, *Bull. Acad. Imper. Sci.*, p. 1800.

65. Garcia, B., Lavallée, S., Perron, G., Michot, C. and Armand, M. (2004). Room temperature molten salts as lithium battery electrolyte, *Electrochim. Acta*, **49**, pp. 4583–4588.

66. Ishikawa, M., Sugimoto, T., Kikuta, M., Ishiko, E. and Kono, M. (2006). Pure ionic liquid electrolytes compatible with a graphitized carbon negative electrode in rechargeable lithium-ion batteries, *J. Power Sources*, **162**, pp. 658–662.

67. Matsumoto, H., Kageyama, H. and Miyazaki, Y. (2002). Room temperature ionic liquids based on small aliphatic ammonium cations and asymmetric amide anions, *Chem. Commun. (Cambridge, UK)*, pp. 1726–1727.

68. Butler, J. A. V. (1936). Hydrogen overvoltage and the reversible hydrogen electrode, *Proc. R. Soc. London, Ser. A*, **157**, pp. 0423–0433.

69. Bard, A. J. and Faulkner, L. R. (2000). *Electrochemical Methods: Fundamentals and Applications* (John Wiley and Sons, New York).

70. Raistrick, I. D. (1983). Lithium insertion reactions in tungsten and vanadium-oxide bronzes, *Solid State Ionics*, **9–10**, pp. 425–430.

71. Bruce, P. G. and Saidi, M. Y. (1992). A 2-step model of intercalation, *Solid State Ionics*, **51**, pp. 187–190.

72. Bruce, P. G. and Saidi, M. Y. (1992). The mechanism of electrointercalation, *J. Electroanal. Chem.*, **322**, pp. 93–105.

73. Kobayashi, S. and Uchimoto, Y. (2005). Lithium ion phase-transfer reaction at the interface between the lithium manganese oxide electrode and the nonaqueous electrolyte, *J. Phys. Chem. B*, **109**, pp. 13322–13326.

74. Fukutsuka, T., Abe, T., Inaba, M. and Ogumi, Z. (2000). Electrochemical intercalation of Li into carbon thin films prepared by plasma CVD, *Mol. Cryst. Liq. Cryst.*, **340**, pp. 517–522.

75. Doi, T., Miyatake, K., Iriyama, Y., Abe, T., Ogumi, Z. and Nishizawa, T. (2004). Lithium-ion transfer at an electrolyte/non-graphitizable carbon electrode interface, *Carbon*, **42**, pp. 3183–3187.

76. Yamada, I., Abe, T., Iriyama, Y. and Ogumi, Z. (2003). Lithium-ion transfer at $LiMn_2O_4$ thin film electrode prepared by pulsed laser deposition, *Electrochem. Commun.*, **5**, pp. 502–505.

Chapter 2

Advanced Lithium and Lithium-Ion Batteries

Martin Finsterbusch and Chih-Long Tsai
Forschungszentrum Jülich, Germany Institute of Energy and Climate Research-Materials Synthesis and Processing (IEK-1), Wilhelm-Johnen-Straße, 52425 Jülich, Germany
m.finsterbusch@fz-juelich.de

2.1 Introduction

The current Li-ion technology has made incredible advances since being introduced in 1991 by Sony. Especially safety, capacity, and lifetime were greatly improved. However, it struggles to keep pace with the increasing needs of modern technology. In addition to more computing power and versatility of consumer electronics, the challenges of the electrification of personal mobility and increasing importance of renewable energy sources in the electric grid demand even higher capacities, power, safety, and lifetime.

Lithium-Ion Batteries: Overview, Simulation, and Diagnostics
Edited by Yoshiaki Kato, Zenpachi Ogumi, and José Manuel Perlado Martin

ISBN 978-981-4800-40-2 (Hardback), 978-0-429-25934-0 (eBook)
www.panstanford.com

As it looks so far, these needs cannot be met with the conventional Li-ion battery technologies, described in more detail in Chapter 1. Advanced cell concepts and chemistries need to be researched and developed to enable another paradigm shift in the battery market. This is especially important as battery performance can enable or impede new technological or societal trends, as can be currently seen in the case of fully electric cars, which suffer from low consumer acceptance due to short driving ranges and in most cases an inconveniently long charging time. However, for an economic breakthrough of any advanced Li or Li-ion technology, it will have to compete in performance and price with the standard Li-ion batteries (and of course other chemistries). As mentioned above, there are three main sectors of battery application with regard to total revenue, each having unique factors for commercial success: (i) consumer electronics, (ii) automotive/mobility, and (iii) stationary (grid) storage.

As for consumer electronics, increased capacity and capacity retention (cyclability) are the main areas of competition against conventional Li-ion battery technologies. Assuming that all other specifications are at least equal, some of the concepts discussed on the following pages either have high theoretical capacities/energy densities or promise excellent cyclability, but no concept was able to demonstrate both yet.

In mobility and especially automotive applications, energy density, price, and safety play an important role. In current fully electric vehicles (EVs), the cost for the battery amounts to a significant part of the total cost of the car, without any reliable experience of its depreciation. Even worse, conventional state-of-the-art Li-ion technology only allows for driving distances two to four times shorter than conventional gas-powered internal combustion engines (ICEs) in standard-sized cars. These facts put the research focus for automotive applications on increased energy density and decreased price per Wh/L stored. Special care needs to be taken to allow for reasonable power density and high C rate charging, since "refueling" time is also a major point for selling EVs. Finally, safety concerns are important, especially in aviation applications, and need to be addressed by advanced battery technology as well.

For stationary applications, the only factors of concern are price per Wh stored and lifetime (cyclability). Most other factors, like energy density, specific energy, and safety, are rather unimportant or can be addressed by proper engineering of the storage site. Due to this fact, other concepts, like redox-flow batteries, are also viable, which would be uneconomical in other areas.

In all three battery application areas, one or more of the following three advanced Li and Li-ion battery technologies might be feasible and promise rather large improvements over existing technologies. The three concepts are lithium-air (Li-air), lithium-sulfur (Li-S), and all-solid-state batteries (ASBs) and will be explained in more detail in this chapter. The focus, however, is on the last one since it (or rather the solid Li-ion conductors used) can also act as an enabler for the first two.

For these three fields of battery application, various roadmaps exist defining research and development targets that are crucial for commercialization of advanced Li/Li-ion concepts. Important roadmaps are regularly issued by the Fraunhofer Institute of System and Innovation Research in Germany (Fig. 2.1a) [1a] and the New Energy and Industrial Technology Development Organization in Japan (Fig. 2.1b) [1b]. All roadmaps define mid- and long-term goals for battery performance parameters and link them to enabling technologies, which are crucial for reaching these goals. For example, high-voltage cathode materials (like a 5 V spinel) will have a large impact on energy density increase in conventional Li-ion batteries, as discussed in Chapter 1, whereas Li-S chemistry promises higher energy density at much lower prices. The roadmaps also give directions to critical aspects of advanced battery technologies and thus where research might be necessary to raise their competitiveness. In the case of ASBs, for example, they are extremely safe but lag in power density. Research on increasing the rate capability of ASBs is thus a major aspect of current research in this field.

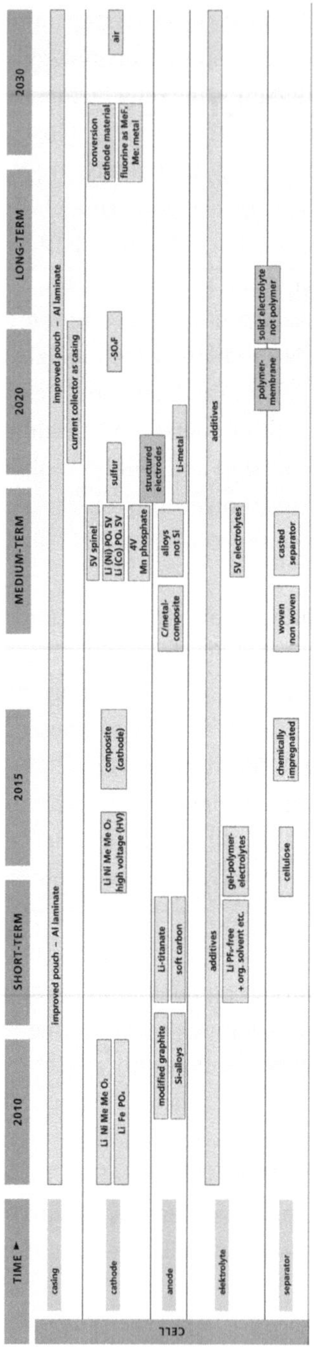
TIME ▸
2010
SHORT-TERM
2015
MEDIUM-TERM
2020
LONG-TERM
2030
CELL
casing
improved pouch – Al laminate
current collector as casing
improved pouch – Al laminate
cathode
Li Ni Me Me O2
Li Fe PO4
Li Ni Me Me O2 high voltage (HV)
composite (cathode)
5V spinel
Li (Ni) PO4 5V
Li (Co) PO4 5V
4V Mn phosphate
sulfur
-SO4F
conversion cathode material
fluorine as MeFx Me: metal
air
anode
modified graphite
Si-alloys
Li-titanate
soft carbon
C/metal-composite
alloys not Si
structured electrodes
Li-metal
elektrolyte
additives
Li PF6-free + org. solvent etc.
gel-polymer-electrolytes
5V electrolytes
additives
separator
cellulose
chemically impregnated
woven non woven
casted separator
polymer-membrane
solid electrolyte not polymer

(a)

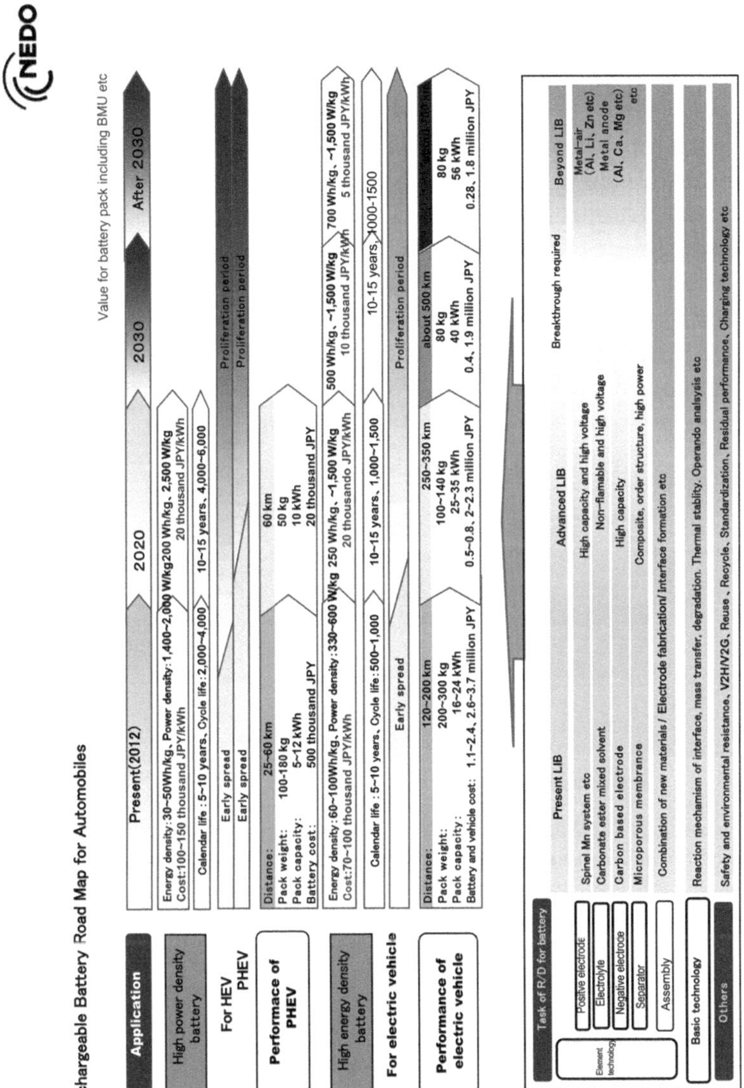

(b)

Figure 2.1 (a) Outtake of the Li battery roadmap issued by the Fraunhofer Institute of System and Innovation Research (ISI) [1] for the cell level only. (b) Full Li battery roadmap (BatteryRM2013) issued by the New Energy and Industrial Technology Development Organization (NEDO) in 2013. (Translated from the Japanese version available at http://www.nedo.go.jp/content/100535728.pdf).

2.1.1 Li-Air Batteries

Li-air batteries are commonly regarded as the ultimate goal in Li-based next-generation battery research and development as the technology promises a huge increase in driving range if used in electrical vehicles. Even though Li-air chemistry was investigated for automotive application in the 1960s and 1970s [2], research stopped due to severe issues with the metal anode and the air cathode. It was not until the mid-1990s, after Abraham and Jiang [3] demonstrated rechargeability of a Li-oxygen cell, that research efforts increased again. However, it is still controversial among researchers whether and how long it will take to realize commercial applications of this system.

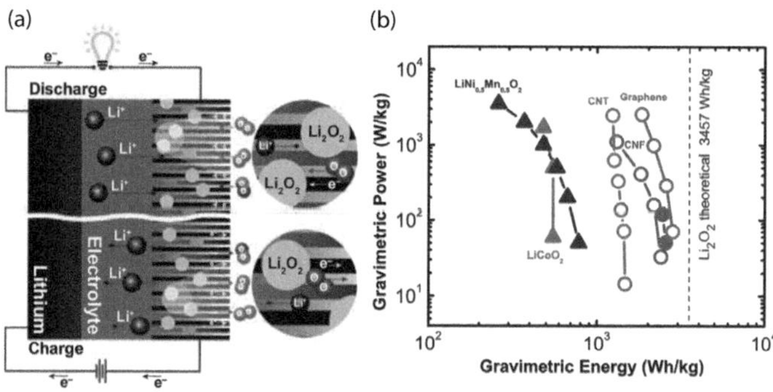

Figure 2.2 (a) Working principle of a Li-O_2 cell and (b) a gravimetric Ragone plot (power vs. energy density) for various C cathodes versus conventional cathodes in Li-ion batteries. From Ref. [6], reproduced with permission.

There are two major concepts for Li-air batteries, the main difference being the electrolyte used. Aqueous systems have a theoretical energy density of 5.6 kW h/L and 3.45 V versus Li^+/Li forming LiOH at the cathode. Nonaqueous systems can form Li_2O at 2.91 V versus Li^+/Li [4], reaching a theoretical energy density of 10.5 kW h/L and thus approaching the energy density of liquid fuels for ICEs. However, Li_2O would be very stable and recharging the cell would be problematic or uneconomical. Li_2O_2 forms at 2.96 V versus Li^+/Li [5] and can be decomposed more easily, making it the more feasible discharge product, reducing the theoretical energy density down to "only" 8 kW h/L.

In either case, Li-air batteries promise an order of magnitude higher capacity than conventional Li-ion batteries, making them a heavily researched next-generation Li-based battery type. In both systems, metallic Li is used as the anode while the cathode is porous, allowing O_2 (ideally from the atmosphere) to enter or leave (see Fig. 2.2a) [6], while providing electronic conductivity and space for the reaction products during discharge.

2.1.1.1 Aqueous Li-air batteries

With water as the main component of the electrolyte and taking part in the reactions, a soluble discharge product forms:

$$2Li^+ + \tfrac{1}{2}O_2 + H_2O + 2e^- \rightarrow 2LiOH$$

One the one hand, this prevents the formation of electrically resistive layers on the cathode; on the other hand, aqueous electrolytes are rather cheap, environmentally benign, and easy to recycle.

However, the specific capacity of the cathode would be limited to approximately 170 mAh/g due to the solubility limit of LiOH of 5.25 M in standard conditions. This limit can only be overcome if a suitable reservoir is provided to allow precipitates to form upon discharge and dissolution of these precipitates upon charging. Another issue on the cathode side is related to the oxygen uptake. While most cells in the lab are run with pure oxygen, if ambient air is used, a selective membrane is needed. It needs to be permeable to O_2 but prevent N and CO_2 from dissolving in the electrolyte, which would otherwise lead to the formation of $LiNO_3$ and Li_2CO_3, respectively. Additionally, the oxygen dissociation reaction at room temperature will require a catalyst, with Pt being the best but most expensive choice if a nonalkaline electrolyte is used [7, 8].

Suitable electrolyte, cathode, and catalyst compositions as well as protective coatings of the Li metal anode are focal points of research in this field. For protective coatings, solid-state Li-ion conductors may be a possible solution and will be discussed later in this chapter.

2.1.1.2 Nonaqueous Li-air batteries

In 1996 the first Li-air cell based on a nonaqueous (aprotic) electrolyte was reported by Abraham and Jiang [9], drawing a lot of attention from the community since finding a suitable electrolyte for reversible cell operation is the main challenge for this battery type.

As the standard organic carbonate–based electrolytes are consumed as the cell is cycled, the formation of Li_2O_2 could not been verified and research on them was mostly abandoned [10–13]. Possible alternatives may be acetonitrile, ionic-liquid, or ether-based electrolytes. The ether-based ones seem to be the most stable and while some Li_2O_2 formation was observed [14, 15] other groups found evidence of decomposition [16].

On the cathode side, issues similar to the ones for aqueous systems exist. Catalysts are needed for the oxygen reduction and especially the oxygen evolution reaction. For example, α-MnO_2 nanowires were able to reduce the polarization resistance and thus increase the coulombic efficiency of the cells to approximately 70% [17]. Catalysts also determine the power density of the cell, since the oxygen reduction and evolution reactions are the rate-limiting steps. So far, power densities of Li-air cells with various catalysts are significantly lower compared to conventional cells. Pore size distribution is also of major concern, since small pores will get blocked easily by the precipitations of the discharge product Li_2O_2 while large pores or catalysts particles provide only a small active area for the reactions. Thus, controlling pore size and matching it according to the Li_2O_2 precipitates enable high-capacity cathodes [18, 19].

Like in aqueous systems, an oxygen-selective membrane is needed on the cathode side if ambient air is to prevent cell degradation by dissolved H_2O or CO_2.

In general, for both types, challenges that need to be addressed by researchers are the electrolytes in general and the optimization of the cathode and the catalyst (mainly porosity and distribution) while the used cell design needs to be tailored to the required rates. For aprotic electrolytes, the oxygen membrane is a crucial point, while for aqueous systems protection of the Li metal from the electrolyte itself is necessary.

2.1.2 Li-S Batteries

To avoid the manifold issues associated with the use of air at the cathode side while still retaining an increase in the theoretical energy density by a factor of 5–10 over conventional Li-ion batteries, the use of sulfur as a cathode material is a viable option. The discharge reaction until the formation of Li_2S yields a theoretical specific energy of 2600 Wh/kg and an energy density of 2800 Wh/L at an average voltage of 2.15 V versus Li^+/Li [20, 21].

$$S_8 + 16Li^+ + 16e^- \rightarrow 8Li_2S$$

In addition to its high capacity, the abundance and thus low cost of sulfur is advantageous. However, Li-S cells suffer from a poor utilization of the cathode material due to the insulating nature of pure sulfur, as well as the final discharge product Li_2S. Furthermore, large capacity fading and poor cyclability are observed due to (i) volume expansion upon lithiation (up to 80%) and resulting pulverization of the sulfur on the cathode side and (ii) the generation of soluble, long-chained polysulfides, which are reduced to Li_2S at the anode, passivating it. Finally, the coulombic efficiency is severely reduced since the long-chained polysulfides can also be reduced to shorter chain lengths and "shuttle" back to the cathode, repeating the cycle [22].

Prevention of this "shuttle mechanism" (Fig. 2.3a) [23] is the focal point for many research groups and, together with S utilization at the cathode, considered the key element for successful commercialization of Li-S cells [22–24].

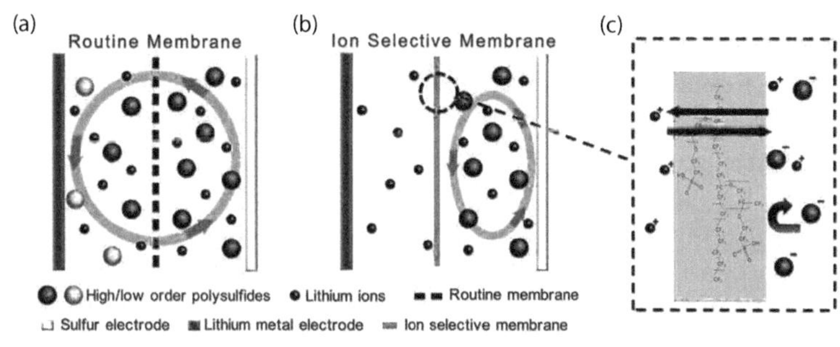

Figure 2.3 Schematic of a standard Li-S cell setup (a), cell with an ion-selective membrane to prevent polysulfide shuttle mechanism (b), and a detailed sketch of the blocking mechanism. Reproduced from Ref. [23] with permission of The Royal Society of Chemistry.

Two strategies are commonly employed. On the one hand, the cathode matrix (e.g., carbon fibers) is tailored to prevent or capture polysulfides while the electrolyte is chosen for low transport of dissolved polysulfides [21, 22, 25, 26]. On the other hand, solid-state Li-ion conductors can be used to inhibit the shuttle mechanism [23, 24]. Again two concepts are conceivable, with the first completely replacing the liquid electrolyte with a solid one, resulting in an ASB, and the second using a thin layer of solid electrolyte as separation between anode and cathode sides (see Fig. 2.3b and 2.3c). ASBs will be discussed in more detail later in this chapter, as well as the use of solid-state Li-ion conductors to advance Li-S battery technology.

Due to their high theoretical energy density, cheap raw materials, and seemingly less severe unresolved issues, Li-S batteries are considered a viable option for mobility and stationary storage applications and could be realized much sooner than Li-air batteries. In fact, several companies, such as Sion Power and Oxis Energy, already have Li-S batteries with high capacity available in their program; however, most of them struggle to compete in lifetime and cyclability with conventional Li-ion technology.

2.2 All-Solid-State Batteries

The third next-generation technology described in this chapter is the ASB. While conventional Li-ion, Li-air, and Li-S batteries have a liquid electrolyte and solid electrodes, the former can be replaced by a solid-state Li-ion conductor. These solid electrolytes, or more precisely the materials used, have a significant impact on possible material choices for the cathode and the anode, as well as cell design, materials processing, and manufacturing of the complete cell.

The biggest advantage of ASBs over conventional batteries is their increased safety. They are inherently leak proof and mechanically more stable, have a higher thermal stability, and are not or less flammable. Also, they often exhibit a much larger electrochemical stability, and no dynamic solid/electrolyte interface (SEI) is formed during the charge/discharge cycles. Thus, improvements in performance can be expected, like higher energy densities at the cell level (high-voltage cathodes and a Li-metal anode) and the system level (less complex battery and system design).

Drawbacks include the more complicated manufacturing process, the often-lower total ionic conductivity of the solid electrolyte, and the solid/solid interfaces, which are prone to delamination/cracking, unwanted phase formation, and many other effects during processing and cycling. They can lead to large interface resistances and thus reduced cell performance or rapid aging. More research is necessary to understand and control these issues and make ASBs competitive with state-of-the art conventional Li-ion batteries.

In addition to these more general advantages and disadvantages, there are specific issues that can be attributed to the two very different main classes of materials. On the one hand, there are organic polymer electrolytes, which might still contain solvents, plasticizers, or ionic liquids but are basically solid in comparison to liquid or polymer-liquid electrolytes used in conventional Li-ion batteries. On the other hand, there are inorganic solid electrolytes, which have much higher thermal, chemical, and mechanical stability and, most importantly, have a larger electrochemical window, enabling the use of high-voltage cathodes and a Li-metal anode.

Both organic polymers and inorganic solid Li-ion conductors often have lower total ionic conductivities than liquid ones; however, the problem is alleviated somewhat for the inorganic materials due to their Li-transfer number of $t^+ = 1$. Due to these facts, this chapter will only focus on inorganic solid-state Li-ion conductors, with the following three sections looking in more detail at the three main materials classes: phosphates, sulfides, and oxides. Here, the material composition, synthesis, and processing and the resulting Li-ion conductivity are examined closely. Section 2.2.4 will then look in more detail at the specifics of ASB cell fabrication and options for promoting their strengths while mitigating their weaknesses.

2.2.1 Phosphates

When NASICON ceramics, with extremely high sodium ion conductivity (**Na** **S**uper**I**onic **CON**ductor), were successfully synthesized, it was found that using Li instead of Na resulted in Li-ion conductors having a good ionic conductivity as well. Nowadays, this class of materials is still commonly referred to as NASICON, even in the context of Li-ion conductors. Their common structure is $LiM_2(PO_4)_3$, and elements for the M site are cations like Ti^{4+}, Ge^{4+}, and Zr^{4+}, with Ti^{4+} being most widely used since it yields the highest Li-ion conductivity in this class of materials. Partial substitution by trivalent cations will increase the Li concentration and the sintering behavior of the material, thus increasing the Li-ion conductivity. $Li_{1+x}Al_xTi_{2-x}(PO_4)_3$ (LATP) has attracted the most attention, with densities around 90% by conventional sintering and total conductivities up to $\sigma_{rt,total} = 7 \times 10^{-4}$ S/cm [27, 28].

When the density is increased further, for example, via spark plasma or hot press sintering techniques, conductivities in the 1×10^{-3} S/cm range can be reached [29].

An easier way to obtain highly conductive LATP is the production of a glass ceramic. While traditional melt quenching or ball milling can be used to produce the glass phase, its conductivity is only in the 10^{-8} S/cm range. Upon heating the material to around 950°C, the crystalline phase precipitates, forming a highly dense microstructure while avoiding the formation of secondary phases at the grain boundaries. Total Li-ion conductivities around $\sigma_{rt,total} = 6 \times 10^{-4}$ S/cm can be reached [30–33], which is at par with conventional sintering of polycrystalline material.

The drawback of Ti at the M site (and other electrolyte materials containing Ti) is the reduction of Ti^{4+} when it comes in contact with Li metal. For example, LATP exhibits a redox peak at 2.4 V versus Li^+/Li due to the reduction of Ti^{4+} to Ti^{3+} [34, 35]. The resulting increase in electronic conductivity makes an application as electrolyte impossible. The only other substitute for Ti^{4+} that has attracted a lot of attention is Ge^{4+}, with $Li_{1.5}Al_{0.5}Ge_{1.5}(PO_4)_3$ (LAGP) having a total Li-ion conductivity of around 6×10^{-4} S/cm [36]. Unfortunately, with Ge being quite expensive, the economic feasibility for widespread application is doubtful.

Pure glass-phase phosphate materials can also be found to be viable as solid-state Li-ion conductors. Li-phosphate-related $Li_xPO_yN_z$, where $x = 2y + 3z - 5$, (LiPON) is one of the best-studied materials in this class of materials, originating and heavily developed at the Oak Ridge National Laboratory, United States [37]. It is stable against Li metal and cathode materials up to 5.5 V. With a total Li-ion conductivity at room temperature of around 1×10^{-6} S/cm, it is still 2 orders of magnitude below crystalline LATP, but due to the cheap, abundant, and nontoxic precursor materials and easy fabrication at room temperature, it is ideal for thin-film ASBs. In fact, thin-film cells based on LiPON are currently the only commercially available all-solid-state Li batteries working at room temperature. These types of cells will be discussed in more detail later in this chapter.

In conclusion, phosphate-based solid-state electrolytes are relatively easy to synthesize (low calcination/sintering temperature and processing in air) and can be rather cheap while having reasonable or even good Li-ion conductivities. However, fabricating highly dense cell components while maintaining good Li-ion conductivity can be challenging.

2.2.2 Sulfides

Like phosphate-based solid electrolytes, sulfide-based crystalline, glass-ceramic, or glass-phase Li-ion conductors are heavily researched within the community and exhibit the highest known values for Li-ion conductivity in a solid so far.

Crystalline sulfide materials with a certain structure are referred to as belonging to the thio-LISICON class, meaning sulfur-substituted **LI**thium **S**uper**I**onic **CON**ductor. Mostly based on Li_4SiS_4 and Li_4GeS_4, Li-ion conductivity can be achieved by substitution of the tetravalent Si or Ge with pentavalent cations, resulting in Li vacancies. Likewise, substitution of Li with divalent cations will lead to reasonable conductivities as well. The highest conductivity is shown by $Li_{3.25}Ge_{0.25}P_{0.75}S_4$ [38], with $\sigma_{rt} = 2.2 \times 10^{-3}$ S/cm, closely followed by $Li_{4.6}Si_{0.4}P_{0.6}S_4$, [39] in the 10^{-4} S/cm range.

In 2012, the crystalline sulfate material $Li_{10}GeP_2S_{12}$ was discovered, having a total Li-ion conductivity at room temperature that even surpasses that of liquid electrolytes. It was first reported by Kamaya et al. [40], and total conductivity was measured at 1.2×10^{-2} S/cm, which is also the highest value found for any solid-state material so far. However, as it contains Ge, of which large-scale use is economically questionable, its commercial application is severely hampered (as is the case for LAGP). Recently, advances have been made by replacing Ge^{4+} with Si^{4+}, however, for conventional synthesis methods, but the resulting conductivities are still lower than those of the Ge compound. Recently, successful preparation of the related $Li_{11}Si_2PS_{12}$ by high-pressure synthesis showed a promising decrease in activation energy and thus largely increased Li-ion conductivity compared to $Li_{10}GeP_2S_{12}$ [41].

Parallel to the crystalline sulfide materials, glass/glass ceramics are heavily researched, also obtaining Li-ion conductivities in the 10^{-3} S/cm range. The most promising materials can be found in the Li_2S–P_2S_5 system, with various ratios of Li_2S and P_2S_5 exhibiting maximum conductivity around a 70:30 ratio. While the purely ceramic material $70Li_2S{\cdot}30P_2S_5$ exhibits a rather low total Li-ion conductivity, in the 10^{-8} S/cm range, the respective glass has conductivity already in the 10^{-5} S/cm range. After partial crystallization of the glass, conductivity always increases for the glass-ceramic material [42], and recently, Seino et al. published a new record of $\sigma_{rt} = 1.2 \times 10^{-2}$ S/cm when the glass-ceramic material was heat-treated at 280°C [43]. As the cause of this large difference between ceramic and glass, a superionic-conducting phase was identified via synchrotron based X-ray powder diffraction. This phase, $Li_7P_3S_{11}$, is metastable and thus not achievable via a solid-state reaction [44].

A further increase in conductivity and chemical stability was achieved by substitution of P_2S_5. An increase by almost a factor of 2 in conductivity (5.4×10^{-3} S/cm) was obtained for $70Li_2S{\cdot}29P_2S_5{\cdot}P_2S_3$, which was identified to have its origin in the S depletion of the superionic phase $Li_7P_3S_{11-y}$ [45]. Other substituents, like SiS_2, GeS_2 TiS_2, and LiI, were also tried, but only LiI led to a significant increase in Li-ion conductivity. However, it also leads to a significant decrease in chemical stability, making it less ideal for application in all-solid-state cells [46].

Another problem for cell manufacturing is the formation of CoS at the electrolyte/cathode interface when using $LiCoO_2$. This effect can, however, be mitigated by proper interface engineering, for example, by application of a thin Co diffusion barrier by coating the cathode particles with a thin Li-permeable oxide layer like TiO_2, $Li_2Ti_2O_5$, or SiO_2 [47, 48].

In conclusion, sulfide-based solid-state Li-ion conductors exhibit outstanding total conductivities and have low synthesis temperatures and a soft texture, which makes ASB fabrication easy. Drawbacks originate from their hygroscopic nature and make material production and cell fabrication rather expensive.

2.2.3 Oxides

In contrast to sulfide-based materials, oxide Li-ion conductors can be easily synthesized and processed in ambient air. However, unlike phosphate-based materials, they need much higher synthesis temperatures. Crystalline materials having perovskite or garnet crystal structures are mainly investigated.

One of the best perovskite-structured solid-state Li-ion conductors is $Li_{3x}La_{(2/3)-x}TiO_3$ (LLT), discovered in 1993 by Inaguma et al. [49]. With bulk conductivities up to 1.2×10^{-3} S/cm at x = 0.11 they would be ideal for all-solid-state cell manufacturing. Unfortunately, the material exhibits a large grain boundary resistance, usually resulting in a much lower total conductivity (e.g., 1.6×10^{-4} S/cm or less) [50, 51].

To solve the issues associated with its high grain boundary resistance, LLT has been heavily investigated for over two decades now, especially with respect to the relationship of crystal structure, Li-ion conduction mechanism, and dopant effects. Two possibilities were found to increase bulk conductivity, but the changes are rather small compared to the changes in undoped material. On the A site, doping with Sr^{2+} led to a lattice expansion and increased the bulk conductivity to $\sigma_{rt,bulk} = 2.54 \times 10^{-3}$ S/cm for $La_{0.53}Sr_{0.03}Li_{0.36}TiO_3$ [52]. On the B site, the substitution of Ti with Al resulted in an increase of $\sigma_{rt,bulk} = 2.95 \times 10^{-3}$ S/cm [52]. But so far, none of the substitutions have been able to increase the grain boundary conductivity.

Like for LATP, a drawback of LLT is the reduction of Ti^{4+} to Ti^{3+} when it comes in contact with Li metal. LLT has a redox peak for Ti^{4+}/Ti^{3+} at approximately 1.8 V versus Li^+/Li, again resulting in high electronic conductivity [53]. Complete substitution of Ti leads to improved chemical stability but results in lower bulk and/or grain boundary conductivity [54].

The discovery of garnet-structured Li-ion conductors having the general formula $Li_xLa_3M_2O_{12}$, with x = 5 to 7 and M = Nb, Ta, Hf, Zr, etc., was made by Thangadurai et al. in 2003 [55]. It was quickly found that the Zr compound $Li_7La_3Zr_2O_{12}$ (LLZ) exhibits the highest total ionic conductivity of approximately 4×10^{-4} S/cm, even though Al doping was found necessary to obtain values this high [56]. The highest conductivity achieved so far for these materials was reported recently by Bernuy-Lopez et al. for Ga-substituted LLZ ($Li_{6.55}Ga_{0.15}La_3Zr_2O_{12}$), with $\sigma_{rt} = 1.35 \times 10^{-3}$ S/cm [57].

In the research community, it is generally accepted that only the cubic LLZ phase yields conductivities this high, while the tetragonal phase has much lower values [58]. Substituents on the Li^+ or Zr^{4+} site were investigated for their impact on stabilizing the cubic phase [56, 59–62]. Al^{3+} on the Li^+ site and Ta^{3+} on the Zr^{4+} site were found to be the most effective. However, it was found recently by the group at Forschungszentrum Jülich that one can obtain high total Li-ion conductivity values for substituted LLZ, even when there is a mixture of cubic and tetragonal phase [63]. Rather than the pure phase composition of the material, the Li content seems to have a much larger effect on the total Li-ion conductivity. It was found that in Al-substituted LLZ the conductivity reached a maximum of 3.4×10^{-4} S/cm (Fig. 2.4) at a Li content of 6.5, despite the 30:70 ratio of the cubic phase to the tetragonal phase (Fig. 2.5). Apparently, at first Al^{3+} got substituted not just at the Li^+ but also at the Zr^{4+} site, leading to a high Li occupancy and leaving only a few vacancies. After sufficient annealing at high temperatures (dwell time in Fig. 2.5), Al^{3+} diffused from the Zr^{4+} to the Li^+ sites, reducing the Li occupancy, creating sufficient vacancies for high ionic conductivity. A similar result was also found for Ta-substituted LLZ by Li et al., who identified the optimal Ta substitution (highest conductivity) for $Li_{7-x}La_3Zr_{2-x}Ta_xO_{12}$ to be $0.4 \le x \le 0.6$, putting the Li content between $6.4 \le Li \le 6.6$ [64–66].

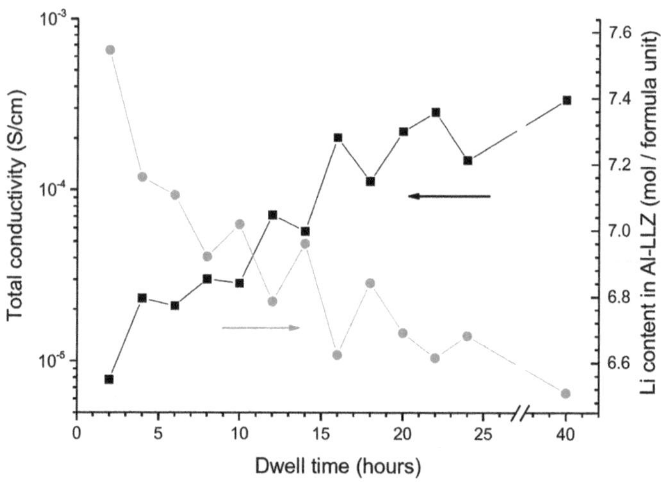

Figure 2.4 Total conductivity and Li content versus sintering time. From Ref. [63]. With permission of Springer.

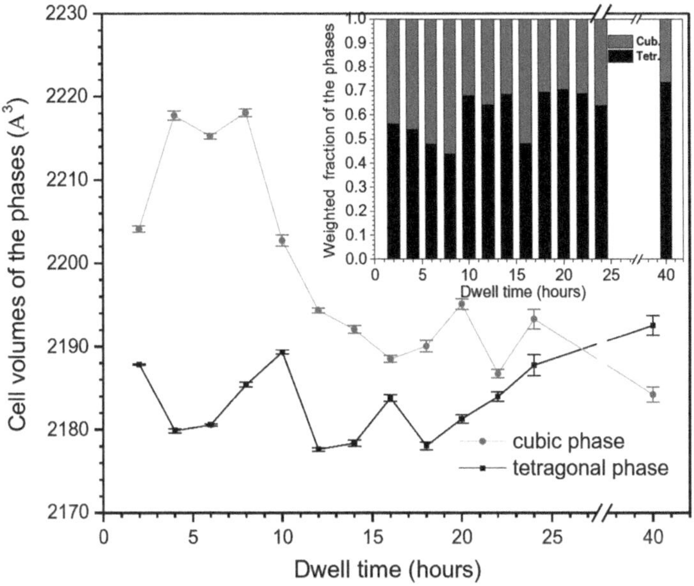

Figure 2.5 Cell volumes of cubic (gray) and tetragonal (black) phase of LLZ:Al versus sintering time. Inset: Fraction of cubic (gray) versus tetragonal (black) phase. From Ref. [63]. With permission of Springer.

In conclusion, the oxide class of solid-state Li-ion conductors has some promising materials with high total Li-ion conductivities. Synthesis and processing in ambient air with conventional ceramic manufacturing techniques would allow a quick translation to industry. On the downside, high synthesis and sintering temperatures, stability issues with Li metal, and high grain boundary resistances are quite challenging for application of these materials and need to be addressed by researchers.

2.2.4 Challenges in Fabrication and Design

Solid-state Li-ion conductors can be incorporated in various ways, either in ASBs or in other advanced Li-batteries, like Li-air or Li-S.

In the case of all-solid-state cells, it was mentioned before that there are cells that are already commercially available. Marketed under the name THINERGY® Micro-Energy Cells by Infinite Power Solutions, Inc., they are based on the thin-film cell design developed at the Oak Ridge National Laboratory. They are normally comprised of a Li metal anode, LiPON as the electrolyte, and $LiCoO_2$ as the cathode, deposited with appropriate current collectors on a substrate for support (Fig. 2.6). With total cell thicknesses around 15 μm and cathode thicknesses around 2 μm, their capacity is very limited.

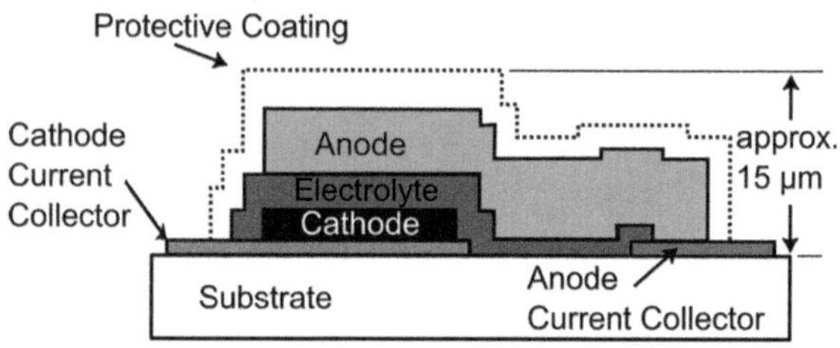

Figure 2.6 Schematic of a thin-film all-solid-state battery [67].

State-of-the-art commercial cells currently have a maximum capacity of 2.2 mAh [68] and are mostly for low-energy applications, like remote sensors and radio-frequency identification (RFID). However, besides the limited capacity, the specifications of these thin-film all-solid-state cells show significant improvement over conventional liquid Li-ion batteries and nicely demonstrate possible gains of this advanced battery technology.

They surpass their conventional "competition" in almost every aspect [68, 69]. For example, operating temperatures from –40°C up to +85°C, high discharge rate capability of over 50 C at only 50% capacity loss, and quick and easy constant voltage charging with 4 C to 90% capacity give a significant edge for automotive application. Furthermore, a low self-discharge and irreversible capacity loss together with an extremely high cycle life (10,000 cycles at 100% depth of discharge with 80% capacity retention) [69] make them ideal for long-term operation.

2.2.4.1 Thin-film all-solid-state cells

Currently in phosphate- and oxide-based thin-film ASB research, two main routes are followed. One of the goals is to increase capacity and/or energy density by tailoring the cathode. Since the Li-ion conductivity of the cathode limits the extractable capacity, materials with larger capacities or higher Li-ion conductivities are researched, as well as cathode materials with a higher voltage versus Li^+/Li. For the latter, new solid-state electrolytes need to be investigated as well, since LiPON is only stable up to 5.5 V. This search can be combined with the second goal, which is further reduction of the internal resistance. This can be achieved via improvement of the Li-ion conductivity of the electrolyte or reduction of the interface resistances in the cell.

For both goals, oxide materials like LLZ would be promising, as they exhibit a high electrochemical stability and Li-ion conductivity almost 2 orders of magnitude higher. However, deposition of a complex, crystalline oxide often requires high substrate temperatures or high-temperature annealing steps, which is problematic for thin films and especially for incorporation into thin-film cell fabrication [70]. To investigate prepared thin-film cells and to connect performance variations to changes in the (micro-) structure or composition, advanced characterization techniques are needed. Since Li cannot be detected with standard X-ray-based analysis techniques, investigation of Li distribution in thin films and identification of secondary phase formation and of grain boundary or interface effects is quite challenging. New methods need to be developed and existing techniques refined to gain deeper insight into solid-state Li-ion conducting materials and their behavior during synthesis and processing. Some important analysis techniques and their application will be explained in the following chapters.

2.2.4.2 Bulk all-solid-state cells

To obtain high capacities, another research direction is the fabrication of all-solid-state cells that have a structure similar to that of conventional Li-ion batteries (from here on referred to as "bulk" ASBs). To overcome the limitation of Li-ion transport in the cathode-active material, it is mixed with the solid-state electrolyte, reproducing the mixed cathodes of conventional Li-ion batteries (which consist of a matrix of porous active material soaked with liquid electrolyte). Since the solid electrolyte is less vulnerable to Li-dendrite growth, no separator is needed and the electrolyte thickness can be reduced to decrease the internal resistance and increase the power density of the cells.

However, if the Li-ion conductivity of the solid-state electrolyte is lower compared to that of conventional liquid electrolytes, cell performance will also be lower if sufficient reduction of electrolyte thickness cannot be achieved (e.g., via appropriate processing). The same argument holds for the mixed cathode and, if applicable, the mixed anode and would result in reduced extractable capacity (limiting the usable electrode thickness). Thus, exploration of highly conductive solid-state electrolytes and their incorporation into cells is still one of the most crucial tasks to enable bulk all-solid-state cells on a commercial level.

Sulfide-based solid-state Li-ion conductors can be easily processed with conventional cathode and anode materials to fabricate an ASB. However, processing has to be done in a protective atmosphere. Mixed-cathode and cell fabrication using oxide-class solid-state Li-ion conductors can be done in ambient air, but material choice can be problematic. They often require synthesis and sintering temperatures above 900°C, where most commonly used cathode materials, like $LiCoO_2$ (LCO), would decompose if processed together. Anode materials like carbon can also not be processed at those temperatures in air. Thus, a lot of research is done (and still necessary) to lower the sintering temperature, for example, by addition of sintering aids like Li_3BO_3 and Al_2O_3 as demonstrated by Ohta et al. [71] for a substituted LLZ:Nb mixed with LCO.

Another very important research area for bulk ASBs is the solid/solid interface between the active material and the electrolyte, since it can have a significant effect on battery performance. Even though no dynamic SEI behavior is observed during charge/discharge cycling, solid-state effects like space charge regions can hinder the Li+ transport across the interface. This problem is augmented when high potentials are introduced, for example, when using high-voltage cathode materials. Understanding of transfer mechanisms at the interface and subsequent interface engineering will thus play an important role in the development of advanced all-solid-state Li and Li-ion batteries. In fact, interface resistance might be the most important issue to make ASBs competitive with conventional LIBs in terms of power density and rate capability [72]. An impressive example of possible gains from proper interface engineering was set by Yada et al. for a 5 V–class cathode material in a Li/LiPON/LNMO ($LiCr_{0.05}Ni_{0.45}Mn_{1.5}O_{4-\delta}$) cell. Upon modification of the interface using $BaTiO_3$ nanoparticles, the charge transfer resistance was reduced by 4 orders of magnitude, the capacity at 0.25 C increased almost threefold (from 45 m Ah/g to 120 m Ah/g), and at 8 C discharge, the cell still retained 84% of its 0.25 C capacity (~100 m Ah/g) [73].

2.2.4.3 Application of solid-state Li-ion conductors and other cell types

As mentioned before, solid-state Li conductors can also enable other advanced Li or Li-ion battery types. If the conductors are chosen properly, their good (electro)chemical stability and their ion selectivity can mitigate or solve some of the innate issues of Li-S and Li-air cells.

For example, the shuttle mechanism in Li-S cells mentioned earlier could be mitigated by the introduction of a solid-state Li-ion conductor. Separation of the anode and cathode sides by an electronically insulating, Li-ion-permeable membrane has proven to largely increase the cycling performance of Li-S cells. Both polymer-based membranes (e.g., Nafion type) [23] and inorganic ceramic sheets (on LATP basis) [24] have already been successfully implemented.

In aqueous Li-air cells, protection of the Li-metal anode from the electrolyte is mandatory. Successful implementation of LATP as anode protection has been shown already [6]. Since LATP is not stable against Li-metal, composite membranes were developed, for example, using LiPON [74] or Li_3N [75].

In conclusion, there are many challenges in the design and fabrication of ASBs. Solid-state Li-ion conductors act as the enabler for ASBs and can also be crucial components in other advanced battery concepts. Intensive research is still necessary to develop new Li-ion conductive materials or tailor existing ones to maximize their performance in the respective application. The promising advantages of ASBs can make research in this area a game changer for the commercialization of advanced Li and Li-ion technologies.

2.3 Outlook

Advanced Li and Li-ion batteries promise a paradigm shift in electromobility, consumer electronics, and stationary storage applications. Energy density, safety, and lifetime can be increased tremendously using new chemistries, materials, or cell designs. Not one technology alone will be able to succeed in all areas, and advanced battery concepts must be tailored to the specific application needs. To that end, for all of the previously described advanced battery technologies, the anode, electrolyte, and cathode as well as the fabrication of complete cells and batteries still need to be addressed by researchers. Specifically, many material science questions need to be solved in order to enable widespread commercial application.

As introduced in this chapter, questions like (electro) chemical stability, electronic and ionic conductivity, composition, microstructure, porosity, and the interface to and reactions with other cell components are typical areas of interest for researchers. However, a simple "as they appear" problem-solving strategy for individual cell components will most likely fail to overcome the complicated and interdependent issues innate to new battery concepts. A holistic approach of materials research, new processing methods, and cell fabrication technology together with simulation and advanced Li diagnostics is needed to enable specific tailoring of all parts of the cell toward a specific application.

Here, the use of Li poses a challenge, especially to processing and diagnostics. High vapor pressures, reactivity, and activity need to be taken into account for processing. For diagnostics, Li location and quantification are difficult or impossible to obtain through standard analysis methods like energy or wavelength dispersive X-ray spectroscopy, X-ray photoelectron spectroscopy, glow discharge optical emission spectroscopy, and secondary ion mass spectrometry. Especially when in situ or nondestructive methods are needed, the mentioned analysis techniques cannot be used. Li-sensitive techniques like neutron scattering (neutron powder diffraction, quasi-elastic neutron scattering, etc.), nuclear reaction analysis, and particle induced X-ray and γ-ray emission spectroscopy are necessary tools to gain insight into the governing mechanisms of advanced Li and Li-ion battery concepts and are discussed in detail in Chapters 4 and 7. Development of these tools and their application in advanced battery research are thus important, and research in this area needs to be pushed forward in parallel to other efforts, for example, in synchrotron-based investigations, detailed in Chapter 4, and combined with advanced modeling of the specific results gained by the various analytical techniques as detailed in Chapter 5. The following chapters of this book are thus dedicated to recent advancements in these crucial analysis methods in the context of conventional and advanced Li and Li-ion battery investigations.

References

1. (a) Technology-Roadmap Lithium-Ion-Batteries 2030, p. 12/13, Dr. Axel Thielmann, Competence Center New Technologies, Fraunhofer-Institute of System- and Innovation Research (ISI), 76139 Karlsruhe Germany; (b) Li battery roadmap issued by New Energy and Industrial Technology Development Organization (NEDO). http://www.nedo.go.jp/content/100535728.pdf.
2. Blurton, K. F. and Sammells, A. F. (1979). Metal/air batteries: their status and potential: a review, *J. Power Sources*, **4**, p. 263.
3. Abraham, K. M. and Jiang, Z. (1996). A polymer electrolyte-based rechargeable lithium/oxygen battery, *J. Electrochem. Soc.*, **143**, p. 1.
4. Chase, Jr., M. W. (1998). NIST-JANAF thermochemical tables (part I and II), *J. Phys. Chem. Ref. Data*, Monograph **9**, p. 1510.
5. Chase, Jr., M.W. (1998). NIST-JANAF thermochemical tables (part I and II), *J. Phys. Chem. Ref. Data*, Monograph **9**, p. 1506.
6. Kwabi, D. G., Ortiz-Vitoriano, N., Freunberger, S. A., Chen, Y., Imanishi, N., Bruce, P. G. and Shao-Horn, Y. (2014). Materials challenges in rechargeable lithium-air batteries, *MRS Bull.*, **39**, pp. 433–452.
7. Giordani, V., Freunberger, S. A., Bruce, P. G., Tarascon, J.-M. and Larcher, D. (2010). H_2O_2 decomposition reaction as selecting tool for catalysts in Li–O_2 cells, *Electrochem. Solid-State Lett.*, **13**, pp. A180–A183.
8. Zhang, T., Imanishi, N., Shimonishi, Y., Hirano, A., Xie, J., Takeda, Y., Yamamoto, O. and Sammes, N. (2010). Stability of a water-stable lithium metal anode for a lithium–air battery with acetic acid–water solutions, *J. Electrochem. Soc.*, **157**, p. A214.
9. Abraham, K. M. and Jiang, Z. (1996). A polymer electrolyte-based rechargeable lithium/oxygen battery, *J. Electrochem. Soc.*, **143**, p. 1
10. Veith, G. M., Nanda, J., Howe, J. Y. and Dudney, N. J. (2011). Spectroscopic characterization of solid discharge products in Li–air cells with aprotic carbonate electrolytes, *J. Phys. Chem. C*, **115**, p. 14325.
11. Freunberger, S. A., Chen, Y., Peng, Z., Griffin, J. M., Hardwick, L. J., Barde, F., Novak, P. and Bruce, P. G. (2011). Reactions in the rechargeable lithium-O_2 battery with alkyl carbonate electrolytes, *J. Am. Chem. Soc.*, **133**, p. 8040.
12. Xu, W., Viswanathan, V. V., Wang, D., Towne, S. A., Xiao, J., Nie, Z., Hu, D. and Zhang, J. G. (2011). Investigation on the charging process of

Li_2O_2-based air electrodes in Li–O_2 batteries with organic carbonate electrolytes, *J. Power Sources*, **196**, p. 3894.

13. Mizuno, F., Nakanishi, S., Kotani, Y., Yokoishi, S. and Iba, H. (2010). Rechargeable Li–air batteries with carbonate-based liquid electrolytes, *Electrochemistry*, **78**, p. 403.
14. Laoire, C. O., Mukerjee, S., Plichta, E. J., Hendrickson, M. A. and Abraham, K. M. (2011). Rechargeable lithium/TEGDME-LiPF6/O2 battery, *J. Electrochem. Soc.*, **158**, p. A302.
15. Freunberger, S. A., Chen, Y., Drewett, N. E., Hardwick, L. J., Bardé, F. and Bruce, P. G. (2011). The lithium–oxygen battery with ether-based electrolytes, *Angew. Chem. Int. Ed.*, **50**, p. 8609
16. Xu, W., Hu, J. and Engelhard, M. (2012). The stability of organic solvents and carbon electrode in nonaqueous Li-O_2 batteries, *J. Power Sources*, **215**, p. 240.
17. Debart, A., Paterson, A. J., Bao, J. and Bruce, P. G. (2008). α-MnO_2 nanowires: a catalyst for the O_2 electrode in rechargeable lithium batteries, *Angew. Chem. Int. Ed.*, **120**, p. 4597.
18. Cheng, H. and Scott, K. (2010). Carbon-supported manganese oxide nanocatalysts for rechargeable lithium–air batteries, *J. Power Sources*, **195**, p. 1370.
19. Yang, X.-H., He, P. and Xia, Y.-Y. (2009). Preparation of mesocellular carbon foam and its application for lithium/oxygen battery, *Electrochem. Commun.*, **11**, p. 1127.
20. Mikhaylik, Y. V. and Akridge, J. R. (2004). Polysulfide shuttle study in the Li/S battery system, *J. Electrochem. Soc.*, **151**(11), pp. A1969–A1976.
21. Ji, X. and Nazar, L. F. (2010). Advances in Li–S batteries, *J. Mater. Chem.*, **20**, pp. 9821–9826.
22. Yang, Y., Zheng, G. and Cui, Y. (2013). Nanostructured sulfur cathodes, *Chem. Soc. Rev.*, **42**, p. 3018.
23. Huang, J.-Q., Zhang, Q., Peng, H.-J., Liu, X.-Y., Qian, W.-Z. and Wei, F. (2014). Ionic shield for polysulfides towards highly-stable lithium–sulfur batteries, *Environ. Sci.*, **7**, p. 347.
24. Yu, X., Bi, Z., Zhao, F. and Manthiram, A. (2015). Hybrid lithium–sulfur batteries with a solid electrolyte membrane and lithium polysulfide catholyte, *ACS Appl. Mater. Interfaces*, **7**, pp. 16625–16631.
25. Ji, X., Lee, K. T. and Nazar, L. F. (2009). A highly ordered nanostructured carbon–sulphur cathode for lithium–sulphur batteries, *Nat. Mater.*, **8**, p. 500.

26. Choi, Y. J., Jung, B. S., Lee, D. J., Jeong, J. H., Kim, K. W., Ahn, H. J., Cho, K. K. and Gu, H. B. (2007.) Electrochemical properties of sulfur electrode containing nano Al_2O_3 for lithium/sulfur cell, *Phys. Scr. T*, **129**, p. 62.
27. Aono, H., Sugimoto, E., Sadaoka, Y., Imanaka, N. and Adachi, G. (1990). Ionic conductivity of solid electrolytes based on lithium titanium phosphate, *J. Electrochem. Soc.*, **137**, p. 1023.
28. Ma, Q., Xu, Q., Tsai, C.-L., Tietz, F. and Guillon, O. (2015). A novel sol–gel method for large-scale production of nanopowders: preparation of $Li_{1.5}Al_{0.5}Ti_{1.5}(PO_4)_3$ as an example, *J. Am. Ceram. Soc.*, **99**, pp. 410–414.
29. Wen, Z., Xu, X. and Li, J. (2009). Preparation, microstructure and electrical properties of $Li_{1.4}Al_{0.4}Ti_{1.6}(PO_4)_3$ nanoceramics, *J. Electroceram.*, **22**, p. 342.
30. Xu, X., Wen, Z., Wu, J. and Yang, X. (2007). Preparation and electrical properties of NASICON-type structured $Li_{1.4}Al_{0.4}Ti_{1.6}(PO_4)_3$ glass-ceramics by the citric acid-assisted sol–gel method, *Solid State Ionics*, **178**, p. 29.
31. Xu, X., Wen, Z., Gu, Z., Xu, X. and Lin, Z. (2004). Lithium ion conductive glass ceramics in the system $Li_{1.4}Al_{0.4}(Ge_{1-x}Ti_x)_{1.6}(PO_4)_3$ (x = 0 – 1.0), *Solid State Ionics*, **171**, p. 207.
32. Narváez-Semanate, J. L. and Rodrigues, A. C. M. (2010). Microstructure and ionic conductivity of $Li_{1+x}Al_xTi_{2-x}(PO_4)_3$ NASICON glass-ceramics, *Solid State Ionics*, **181**, p. 1197.
33. Cruz, A. M., Ferreira, E. B. and Rodrigues, A. C. M. (2009). Controlled crystallization and ionic conductivity of a nanostructured $LiAlGePO_4$ glass–ceramic, *J. Non-Cryst. Solids*, **355**, p. 2295.
34. Birke, P., Salam, F., Döring, S. and Weppner, W. (1999). A first approach to a monolithic all solid state inorganic lithium battery, *Solid State Ionics*, **118**, p. 149.
35. Wu, X. M., Li, X. H., Wang, S. W., Wang, Z., Zhang, Y. H., Xu, M. F. and He, Z. Q. (2003). Preparation and characterization of lithium-ion-conductive $Li_{1.3}Al_{0.3}Ti_{1.7}(PO_4)_3$ thin films by the solution deposition, *Thin Solid Films*, **425**, p. 103.
36. He, K., Wang, Y., Zu, C., Zhao, H., Liu, Y., Chen, J., Han, B. and Ma, J. (2011) Structural analysis of nickel doped cobalt ferrite nanoparticles prepared by coprecipitation route, *Physica B*, **406**, p. 3947.
37. Bates, J. B., Dudney, N. J., Neudecker, B., Ueda, A. and Evans, C. D. (2000). Thin-film lithium and lithium-ion batteries, *Solid State Ionics*, **135**, p. 33.
38. Kanno, R. and Murayama, M. (2001). Lithium ionic conductor thio-LISICON: the $Li_2SGeS_2P_2S_5$ system, *J. Electrochem. Soc.*, **148**, p. A742.
39. Murayama, M., Kanno, R., Irie, M., Ito, S., Hata, T., Sonoyama, N. and Kawamotow, Y. (2002). Synthesis of new lithium ionic conductor thio-

LISICON: lithium silicon sulfides system, *J. Solid State Chem.*, **168**, p. 140.

40. Kamaya, N., Homma, K., Yamakawa, Y., Hirayama, M., Kanno, R., Yonemura, M., Kamiyama, T., Kato, Y., Hama, S., Kawamoto, K. and Mitsui, A. (2011). A lithium superionic conductor, *Nat. Mater.*, **10**, p. 682.
41. Kuhn, A., Gerbig, O., Zhu, C., Falkenberg, F., Maier, J. and Lotsch, B. V. (2014). A new ultrafast superionic Li-conductor: ion dynamics in $Li_{11}Si_2PS_{12}$ and comparison with other tetragonal LGPS-type electrolytes, *Phys. Chem. Chem. Phys.*, **16**, pp. 14669–14674.
42. Tatsumisago, M. and Hayashi, A. (2008). Preparation of lithium ion conducting glasses and glass–ceramics for all-solid-state batteries, *J. Non-Cryst. Solids*, **354**, p. 1411.
43. Seino, Y., Ota, T., Takada, K., Hayashi, A. and Tatsumisago, M. (2014). A sulphide lithium super ion conductor is superior to liquid ion conductors for use in rechargeable batteries, *Energy Environ. Sci.*, **7**, p. 627.
44. Mizuno, F., Hayashi, A., Tadanaga, K. and Tatsumisago, M. (2005). New, highly ion-conductive crystals precipitated from $Li_2S–P_2S_5$ glasses, *Adv. Mater.*, **17**, p. 918.
45. Minami, K., Hayashi, A., Ujiie, S. and Tatsumisago, M. (2011). Electrical and electrochemical properties of glass–ceramic electrolytes in the systems $Li_2S–P_2S_5–P_2S_3$ and $Li_2S–P_2S_5–P_2O_5$, *Solid State Ionics*, **192**, p. 122.
46. Kim, Y. and Martin, S. W. (2006). Ionic conductivities of various GeS_2-based oxy-sulfide amorphous materials prepared by melt-quenching and mechanical milling method, *Solid State Ionics*, **177**, p. 2881.
47. Sakuda, A., Hayashi, A. and Tatsumisago, M. (2010). Electrochemical performance of all-solid-state lithium secondary batteries improved by the coating of $Li_2O–TiO_2$ films on $LiCoO_2$ electrode, *J. Power Sources*, **195**, p. 599.
48. Sakuda, A., Kitaura, H., Hayashi, A., Tadanaga, K. and Tatsumisago, M. (2009). All-solid-state lithium secondary batteries with oxide-coated $LiCoO_2$ electrode and $Li_2S–P_2S_5$ electrolyte, *J. Power Sources*, **189**, p. 527.
49. Inaguma, Y., Chen, L., Itoh, M., Nakamura, T., Uchida, T., Ikuta, H. and Wakihara, M. (1993). High ionic conductivity in lithium lanthanum titanate, *Solid State Commun.*, **86**, p. 689.
50. Kawai, H. and Kuwano, J. (1994). Lithium ion conductivity of A-site deficient perovskite solid solution $La_{0.67-x}Li_{3x}TiO_3$, *J. Electrochem. Soc.*, **L78**, p. 141.

51. Geng, H., Lan, J., Mei, A., Lin, Y. and Nan, C. W. (2011). Effect of sintering temperature on microstructure and transport properties of $Li_{3x}La_{2/3-x}TiO_3$ with different lithium contents, *Electrochim. Acta*, **56**, p. 3406.

52. Morata-Orrantia, A., García-Martín, S. and Alario-Franco, M. Á. (2003). New $La_{2/3-x}Sr_xLi_xTiO_3$ solid solution: structure, microstructure, and Li+ conductivity, *Chem. Mater.*, **15**, p. 363.

53. Chen, C. H., Xie, S., Sperling, E., Yang, A. S., Henriksen, G. and Amine, K. (2004). Stable lithium-ion conducting perovskite lithium–strontium–tantalum–zirconium–oxide system, *Solid State Ionics*, **167**, p. 263.

54. Inaguma, Y., Chen, L., Itoh, M. and Nakamura, T. (1994). Candidate compounds with perovskite structure for high lithium ionic conductivity, *Solid State Ionics*, **70/71**, pp. 196–202.

55. Thangadurai, V., Kaack, H. and Weppner, W. J. F. (2003). Novel fast lithium ion conduction in garnet-type $Li_5La_3M_2O_{12}$ (M = Nb, Ta), *J. Am. Ceram. Soc.*, **86**, p. 437.

56. Buschmann, H., Dölle, J., Berendts, S., Kuhn, A., Bottke, P., Wilkening, M., Heitjans, P., Senyshyn, A., Ehrenberg, H., Lotnyk, A., Duppel, V., Kienle, L. and Janek, J. (2011). Structure and dynamics of the fast lithium ion conductor "$Li_7La_3Zr_2O_{12}$", *Phys. Chem. Chem. Phys.*, **13**, p. 19378.

57. Bernuy-Lopez, C., Manalastas, W., Jr., Del Amo, J. M. L., Aguadero, A., Aguesse, F. and Kilner, J. A. (2014). Atmosphere controlled processing of Ga-substituted garnets for high Li-ion conductivity ceramics, *Chem. Mater.*, **26**, pp. 3610–3617.

58. Awaka, J., Kijima, N., Hayakawa, H. and Akimoto, J. (2009). Synthesis and structure analysis of tetragonal $Li_7La_3Zr_2O_{12}$ with the garnet-related type structure, *J. Solid State Chem.*, **182**, p. 2046.

59. Tan, J. and Tiwari, A. (2012). Synthesis of cubic phase $Li_7La_3Zr_2O_{12}$ electrolyte for solid-state lithium-ion batteries, *Electrochem. Solid-State Lett.*, **15**(3), p. A37.

60. O'Callaghan, M. P., Powell, A. S., Titman, J. J., Chen, G. Z. and Cussen, E. J. (2008). Switching on fast lithium ion conductivity in garnets: the structure and transport properties of $Li_{3+x}Nd_3Te_{2-x}Sb_xO_{12}$, *Chem. Mater.*, **20**, p. 2360.

61. Xie, H., Li, Y. and Goodenough, J. B. (2012). Low-temperature synthesis of $Li_7La_3Zr_2O_{12}$ with cubic garnet-type structure, *Mater. Res. Bull.*, **47**, p. 1229.

62. Kotobuki, M., Kanamura, K., Sato, Y. and Yoshida, T. (2011). Fabrication of all-solid-state lithium battery with lithium metal anode using Al_2O_3-added $Li_7La_3Zr_2O_{12}$ solid electrolyte, *J. Power Sources*, **196**, p. 7750.

63. Tsai, C. L., Dashjav, E., Hammer, E.-M., Finsterbusch, M., Tietz, F., Uhlenbruck, S. and Buchkremer, H. P. (2015). High conductivity of mixed phase Al-substituted $Li_7La_3Zr_2O_{12}$, *J. Electroceram.*, **35**, pp. 25–32.
64. Li, Y., Han, J.-T., Wang, C.-A., Vogel, S. C., Xie, H., Xu, M. and Goodenough, J. B. (2012). Ionic distribution and conductivity in lithium garnet $Li_7La_3Zr_2O_{12}$, *J. Power Sources*, **209**, p. 278.
65. Li, Y., Han, J.-T., Wang, C.-A., Xie, H. and Goodenough, J. B. (2012). Optimizing Li+ conductivity in a garnet framework, *J. Mater. Chem.*, **22**, p. 15357.
66. Xie, H., Alonso, J. A., Li, Y., Fernandez-Díaz, M. T. and Goodenough, J. B. (2011). Lithium distribution in aluminum-free cubic $Li_7La_3Zr_2O_{12}$, *Chem. Mater.*, **23**, p. 3587.
67. Bates, J. B., Dudney, N. J., Neudecker, B., Ueda, A. and Evans, C. D. (2000). Thin-film lithium and lithium-ion batteries, *Solid State Ionics*, **135**, p. 33.
68. http://www.cytech.com/products-ips
69. http://www.infinitepowersolutions.com/images/stories/downloads/ds1015_MEC202_v1-2FINALDRAFT_20120711.pdf
70. Lobe, S., Dellen, C., Finsterbusch, M., Gehrke, H.-G., Sebold, D., Tsai, C.-L., Uhlenbruck, S. and Guillon, O. (2016). Radio frequency magnetron sputtering of $Li_7La_3Zr_2O_{12}$ thin films for solid-state batteries, *J. Power Sources*, **307**, pp. 684–689.
71. Ohta, S., Seki, J., Yagi, Y., Kihira, Y., Tani, T. and Asaoka, T. (2014). Co-sinterable lithium garnet-type oxide electrolyte with cathode for all-solid-state lithium ion battery, *J. Power Sources*, **265**, pp. 40–44.
72. Sakuda, A., Hayashi, A. and Tatsumisago, M. (2010). Electrochemical performance of all-solid-state lithium secondary batteries improved by the coating of Li_2O–TiO_2 films on $LiCoO_2$ electrode, *J. Power Sources*, **195**, pp. 599–603.
73. Yada, C., Ohmori, A., Ide, K., Yamasaki, H., Kato, T., Saito, T., Sagane, F. and Iriyama, Y. (2014). Dielectric modification of 5V-class cathodes for high-voltage all-solid-state lithium batteries, *Adv. Energy Mater.*, **4**, p. 1301416.
74. Stevens, P., Toussaint, G., Caillon, G., Viaud, P., Vinatier, P., Cantau, C., Fichet, O., Sarrazin, C. and Mallouki, M. (2010). Development of a lithium air rechargeable battery, *ECS Trans.*, **28**, p. 1.
75. Visco, S., Katz, B., Chu, M. Y. and De Jonghe, L., eds. (2009). *Symposium on Scalable Energy Storage: Beyond Lithium-Ion* (IBM, Almaden Institute, San Jose, CA).

Chapter 3

Advanced Diagnostics of Lithium-Ion Batteries

Yuki Orikasa[a] and Yoshiharu Uchimoto[b]
[a]*Department of Applied Chemistry, Ritsumeikan University, 1-1-1 Nojihigashi, Kusatsu, Shiga 525-8577, Japan*
[b]*Graduate School of Human and Environmental Studies, Kyoto University, Yoshidahonmatsucho, Sakyoku, Kyoto 606-8501, Japan*
uchimoto.yoshiharu.2n@kyoto-u.ac.jp

3.1 Introduction

Power storage devices, typified by lithium-ion secondary batteries, are high-efficiency devices for electrochemical energy conversion. The use of batteries in automobiles and smart grids is currently widespread, and it is hoped that the adoption of power storage devices in these contexts will play a large part in resolving environmental and energy issues in the future [1]. To improve the performance of these devices and achieve high-performance storage batteries, it is required to understand the successive hierarchical reactions that occur within these devices, over a wide spatial and temporal range. The reactions that occur inside lithium-ion secondary batteries involve multiple inter-related phenomena with nonequilibrium

Lithium-Ion Batteries: Overview, Simulation, and Diagnostics
Edited by Yoshiaki Kato, Zenpachi Ogumi, and José Manuel Perlado Martin

ISBN 978-981-4800-40-2 (Hardback), 978-0-429-25934-0 (eBook)
www.panstanford.com

condition. Therefore, we cannot at present accurately grasp the factors governing reaction kinetics or the mechanism of degradation by using the destructive analyses that have conventionally been employed in storage battery research.

The mechanisms behind reactions in a lithium-ion secondary battery at their respective spatial and temporal scales are shown in Fig. 3.1. The lithium-ion secondary battery at the level we generally use is located in the top right, where the scale is the largest. It is composed of layered cathodes, anodes, and electrolyte-containing separators. The electrode layers possess a complex 3D structure, with a mixture of active material, conductive carbon, and binding agent used in the cathodes and anodes. The reaction proceeds by repeated insertion/extraction of lithium ions into/from the active material. However, ion diffusion causes lattice reorganization within the active material and phase changes to occur along with it. Lithium-ion insertion/extraction occurs in association with the charge-transfer reaction at the active material/electrolyte interface, forming a reaction site.

Lithium-ion secondary batteries are on the order of centimeters at the pack level. Composite electrodes, meanwhile, range from a few tens of micrometers to one hundred micrometers thick. The active material, on the other hand, is on the order of tens of nanometers to micrometers. Thus, the reaction at the electrode/electrolyte interface proceeds in an interfacial phase formed on the order of nanometers. As with the spatial scales, the temporal scales required for the reaction cover a wide range, from hours to milliseconds. Therefore, precisely ascertaining the spatiotemporal hierarchical structure of lithium-ion secondary batteries is essential if we are to enhance their performance.

However, their sealed structure means visualizing these devices is not easy, and so the influence of the above-mentioned hierarchical reaction structure on battery characteristics has never been clarified. To elucidate the mechanisms behind rate-limiting reactions and instability phenomena, research and development are needed for new, advanced analytic technology that can observe electrochemical reactions in operating conditions. Determining the limitations of existing devices, as well as their respective causes, is likewise essential for the development of component technology for innovative storage batteries.

Therefore, in the present study, we developed a novel *operando* measurement technique, suitable for various reaction scales, that utilizes synchrotron X-ray radiation, taking advantage of its strong transmission power and high spatiotemporal resolution. We then made direct observations of the reactions that occur in storage batteries under operating conditions and elucidated their reaction mechanisms. Our results could lead to the creation of lithium-ion secondary batteries with higher cyclability and higher power densities, as well as to other, innovative storage batteries.

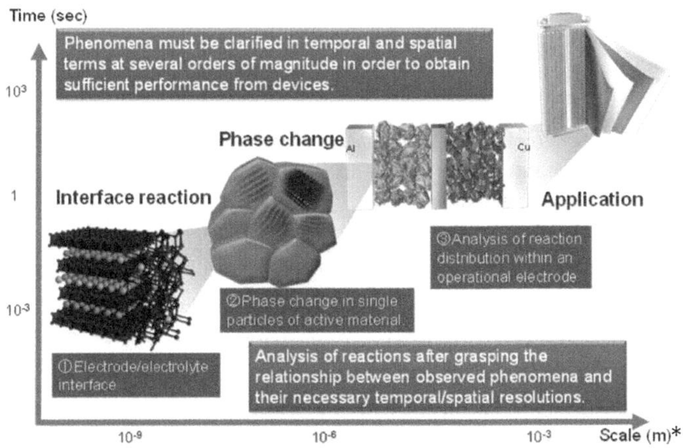

Figure 3.1 Schematic diagram of reactions in lithium-ion secondary batteries at various spatial and temporal scales.

3.2 Analysis of the Reaction Mechanism at the Electrode/Electrolyte Interface

The reaction site in a lithium-ion secondary battery is the electrode/electrolyte interface shown in Fig. 3.2. The difference between the internal potentials of the electrode and the electrolyte establishes a potential difference (i.e., the equilibrium electrode potential) at the electrode/electrolyte interface, leading to charge separation. This potential difference causes a region with a potential gradient, called the electric double layer, to appear on the electrolyte side of the interface. This electric double layer serves as the reaction site, at which ion migration and desolvation and changes in electrode structure occur. On the electrode side, on the other hand, a complex space charge layer is generated, apparently due to the aforementioned potential difference.

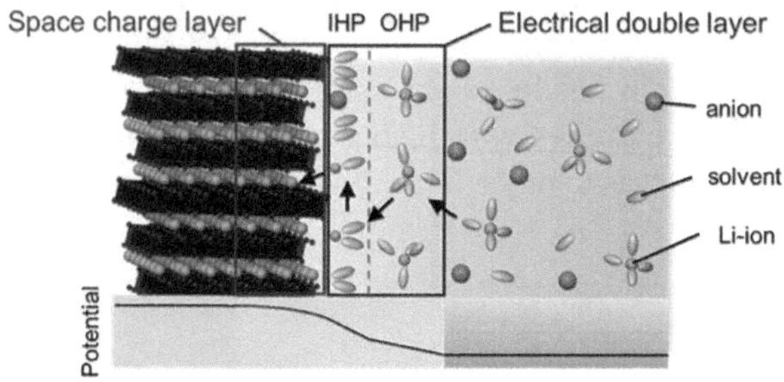

Figure 3.2 Schematic diagram of the electrode/electrolyte interface in a lithium-ion secondary battery. IHP and OHP mean inner Helmholtz plane and outer Helmholtz plane, respectively.

Acquiring information about the electrode/electrolyte interface is extremely important for the future development of storage batteries [2]. Electrochemical impedance spectroscopy methods are relatively simple but powerful tools for elucidating phenomena at the electrode/electrolyte interface. Unfortunately, they cannot give direct information about the specifics of the interfacial structure. Therefore, the development of new methods by which the electrode/electrolyte interface can be directly observed is desired.

Recent years have seen examples of in situ observations of the electrode/electrolyte interface in thin-film model electrodes, made using X-ray reflectivity (XRR) methods [3, 4] and Fourier-transform infrared (FT-IR) spectroscopy methods [5]. Density and roughness information about the electrode surface can be obtained using XRR methods, while information about the organic coatings that form on the electrode surface can be obtained using FT-IR methods. These analysis methods are important for interface research, but many aspects of the electrode/electrolyte interfaces in lithium-ion secondary batteries remain "black boxes." In this paper, therefore, we focus on total reflection fluorescence through X-ray absorption spectroscopy (XAS) employing synchrotron X-ray radiation as a method for *operando* observation of the electrode/electrolyte interface.

Total reflection fluorescence using XAS is a method for detecting X-ray fluorescence emitted from the surface of a specimen under conditions where incident X-rays are totally reflected. Total reflection is a phenomenon whereby, when light from a medium with a high refractive index enters a medium with a low refractive index, all light is reflected without traveling into the material when the incident angle is less than a given angle (the critical angle θ_c). In the visible light region, the refractive index is typically lower for a gas than for a solid or a liquid. Thus, total internal reflection typically occurs when light travels into a gas from a solid or a liquid. In the X-ray wavelength region, on the other hand, the refractive index is lower for liquids and solids than for gases, and so total internal reflection occurs when light travels into a solid or a liquid from a gas.

The critical angle θ_c is formulated using Snell's law and is dependent on the electron density of the material and the energy of the irradiating X-rays. It can be expressed as:

$$\theta_c \approx \frac{20.3\sqrt{\rho}}{E_X}, \qquad (3.1)$$

where θ_c is the critical angle (mrad), ρ is the material density (g/cm^3), and E_X is the X-ray energy (keV) [6]. Assuming the crystal density of $LiCoO_2$ (the specimen used in this study) is ~5.05 g/cm^3, we calculate the critical angle at the *K*-edge absorption edge of Co (~7.71 keV) to be ~5.9 mrad (0.34°).

Under conditions of total reflectance, almost no X-rays penetrate into the material, but a small fraction does travel inside the material as an evanescent wave. Its penetration depth varies according to the X-ray's energy and the material's electron density, but it is largely on the order of a few nanometers. The depth at which evanescent waves can be detected in $LiCoO_2$ in lithium-ion secondary batteries is known to be 2– 5 nm, in the vicinity of the critical angle for total reflection. An evanescent wave has the same energy as the irradiating X-ray: X-ray absorption knocks inner-shell electrons out of atoms in the area penetrated by the evanescent wave, and they emit X-ray fluorescence and Auger electrons.

This emitted X-ray fluorescence can be detected through the use of total reflection fluorescence using XAS. In this study, we devised an *operando* measurement cell in which the conditions necessary for total reflection could be realized. Using a $LiCoO_2$ cathode as a model case, we obtained information about the interface on the cathode side in lithium-ion secondary batteries under operating conditions.

We measured X-ray absorption near-edge structure (XANES) spectra for the prepared $LiCoO_2$ thin-film specimen ("Pre-immersion" in Fig. 3.3) and for the specimen in contact with the electrolyte in the cell ("Post-immersion" in Fig. 3.3). A comparison of the bulk XANES spectra (obtained through normal oblique incidence measurements of thin films) and surface XANES spectra (obtained through measurements at an angle lower than the critical angle for total internal reflection) is shown in Fig. 3.3.

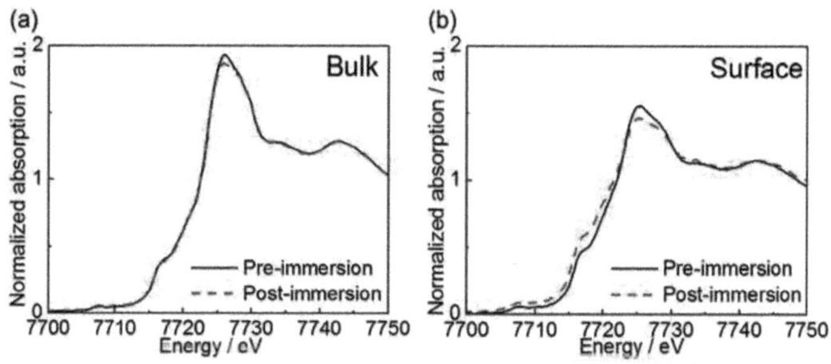

Figure 3.3 Co *K*-edge XANES spectra for $LiCoO_2$ before and after immersion in the electrolyte: (a) bulk data and (b) surface data. Reprinted from Ref. [7] with permission from John Wiley and Sons.

From these results, we can see that the surface spectrum for the $LiCoO_2$/electrolyte interface changes greatly when the electrode comes into contact with the electrolyte, yet the bulk spectrum hardly changes at all. The valence state for Co is greatly reflected in the XANES spectra shown in Fig. 3.3 for the Co *K*-edge. The energy of the absorption edge in the surface spectrum has shifted toward the low-energy side, meaning that Co ions within $LiCoO_2$ near the interface have been reduced by contact with the electrolyte.

The Co ions in $LiCoO_2$ are oxidized in the charging step and reduced in the discharging step. These correspond to high-energy and low-energy shifts of the Co *K*-edge XANES, respectively. We measured XANES spectra at potentials of 3.2 V, 3.8 V, 4.0 V, and 4.2 V for the charging step and at 4.0 V and 3.8 V for the discharging step. We plotted energy values that were intensity normalized by 0.5 (Fig. 3.4).

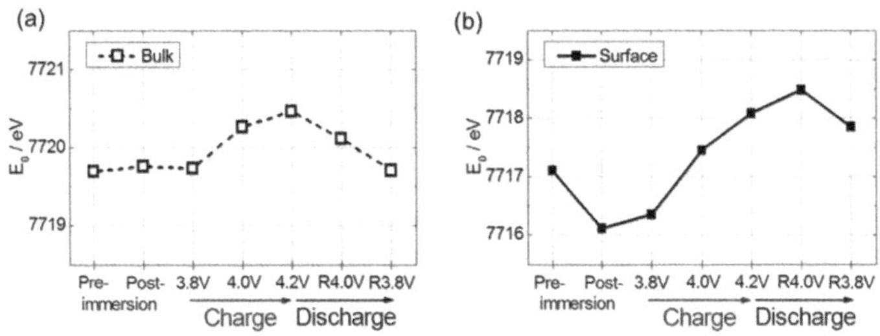

Figure 3.4 Absorption edge energy of (a) bulk and (b) surface Co *K*-edge XANES for $LiCoO_2$ accompanying the charge/discharge reaction. Reprinted from Ref. [7] with permission from John Wiley and Sons.

As shown in Fig. 3.4, the valence state of Co obtained from the bulk XANES spectra changes reversibly following charging and discharging. The valence of Co obtained from the surface XANES spectra, on the other hand, changes irreversibly, unlike the bulk spectra. This demonstrates that irreversible behavior has occurred across the crystal interface. In short, as Fig. 3.5 shows, the initial deterioration that arose from the electrode coming into contact with the electrolyte increases with charging and discharging and appears to cross the interface to the bulk [7]. This phenomenon demonstrates why a stable electrode/electrolyte interface is essential to extending the cycle stability of lithium-ion secondary batteries. We thus decided to attempt to construct a stable interface in this study.

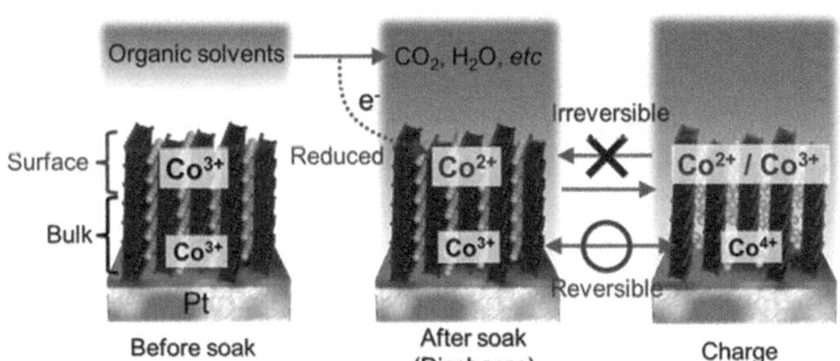

Figure 3.5 Models of the reaction mechanism for the electrode/electrolyte interface for $LiCoO_2$.

We coated the cathode-active material with MgO as a way to control the electrode/electrolyte interface. After preparing a $LiCoO_2$ thin-film model electrode, we coated the entire surface with a few-nanometer-thick layer of MgO. The charge and discharge densities for the noncoated and MgO-coated $LiCoO_2$ electrodes were determined from cyclic voltammograms, and the coulomb efficiency was calculated at each cycle. The interphase ion transfer resistances at the electrode/electrolyte interface at each potential were calculated from AC impedance–based measurements. Both sets of data are shown in Fig. 3.6. The coulomb efficiency of the noncoated electrode decreased remarkably at high potentials, whereas the MgO coating kept the coulomb efficiency high. In addition, the triple-digit increase observed for interphase ion transfer resistance for the noncoated electrode appears to have been suppressed by the MgO coating. The above demonstrates that an MgO surface coating improves stability at high potentials.

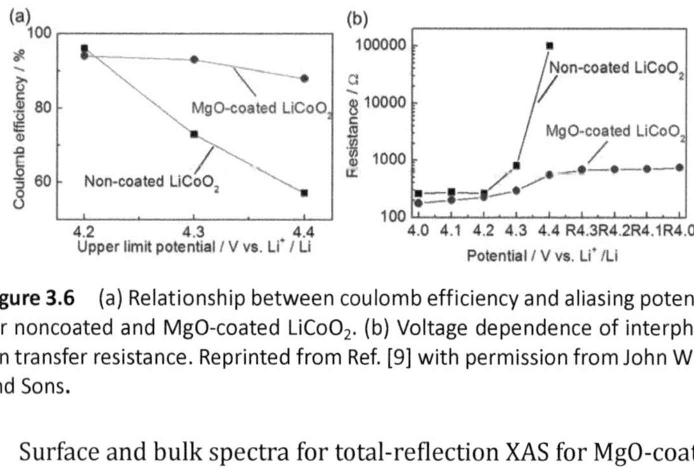

Figure 3.6 (a) Relationship between coulomb efficiency and aliasing potential for noncoated and MgO-coated $LiCoO_2$. (b) Voltage dependence of interphase ion transfer resistance. Reprinted from Ref. [9] with permission from John Wiley and Sons.

Surface and bulk spectra for total-reflection XAS for MgO-coated $LiCoO_2$ are shown in Fig. 3.7. We can see that the bulk spectrum is not changed before and after electrolyte immersion. This suggests that, similar to noncoated $LiCoO_2$, the change occurring at the electrode/electrolyte interface at the point in time when the electrode comes into contact with the electrolyte does not reach the bulk. Moreover, we can see that the location of the absorption edge has not changed in the surface spectrum. We believe that the interfacial layer formed by contact with the electrolyte, as seen for noncoated $LiCoO_2$, has been suppressed here. We conclude that a solid solution phase containing Mg is generated at the interface by the MgO coating. This solid solution plays a pivotal role, stabilizing the interface while keeping the crystal structure stable after lithium ions have been desorbed at high voltages. The MgO coating thus helps to improve the cell's electrochemical properties [8, 9].

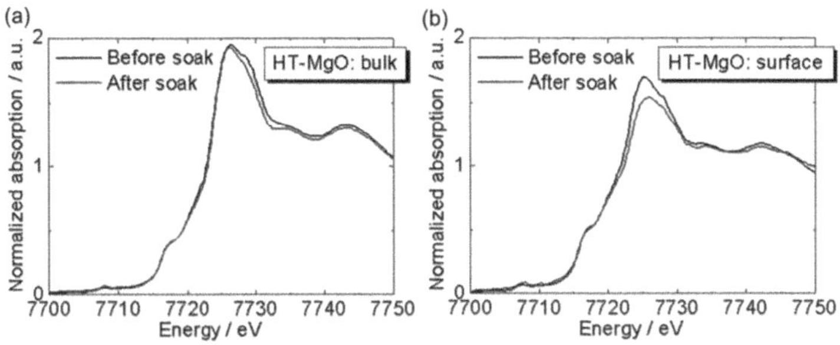

Figure 3.7 XANES spectra for Mg-coated $LiCoO_2$ before and after electrolyte immersion: (a) bulk data and (b) interface data. Reprinted from Ref. [9] with permission from John Wiley and Sons.

3.3 Nonequilibrium Phase Transition Behaviors of the Electrode-Active Material

Following the reaction at the electrode/electrolyte interface, lithium-ion insertion/extraction occurs within the electrode-active material. Under operating conditions, the electrode-active material in a lithium-ion secondary battery undergoes a phase change due to the continuous exchange of ions and electrons. Of particular note, phase change behavior under high-speed charging/discharging conditions may differ from the phase change behavior at equilibrium. Conventional analyses have mainly examined the phase state and electronic state at equilibrium. Thus, even though these nonequilibrium structural changes are directly and essentially connected to device performance, no information on them has yet been obtained.

Thus, in this study, adopting $LiFePO_4$ as a model material, we clarify the phase changes over time in the electrode-active material in a nonequilibrium state and introduce a test case where we attempt to elucidate the origin of the high-speed reaction. $LiFePO_4$ has the advantages of low cost, owing to the relatively inexpensive Fe being used as its principal component, high crystal-structure stability due to its PO_4 backbone, and superior cycling performance and safety [10]. When this material was initially reported, there were issues of insufficient output characteristics due to its low electrical conductivity. However, it has since achieved output characteristics superior to those of existing $LiCoO_2$-type materials, thanks to microparticulation [11] and carbon-coating [12] processes. In charge/discharge reactions, $LiFePO_4$ reportedly phase-separates into a Li-rich $LiFePO_4$ (LFP) phase and a Li-poor $FePO_4$ (FP) phase; the total lithium content varies only in terms of the changing proportions of the two phases [13].

Various models have been devised for the LFP/FP phase transition mechanism in its two-phase coexistence reaction. However, the phase transition behavior of $LiFePO_4$ is uncertain, and new models continue to be proposed even today. One reason for the failure to clearly determine its phase transition behavior is that nearly all of the models proposed to date have merely conjectured about the nonequilibrium state, on the basis of measurements at equilibrium or on calculations. Learning how phases change in real charge/discharge reactions is an essential step for creating design guidelines for new materials; a solution to this problem is eagerly awaited.

In this study, we directly observed a previously unknown phase transition behavior in a two-phase coexistence-type $LiFePO_4$ cathode by applying high-speed time-resolved synchrotron-based measurements. Our results revealed the features of this cathode material, which possesses high-speed charge/discharge characteristics [14–16]. We uncovered the phase behavior under nonequilibrium conditions, a state that has not yet been discussed except in conjecture from calculations and steady-state measurements. This insight led to a general characterization of two-phase coexistence-type electrodes, making this study an important one indeed.

The time-resolved XRD measurements and charging curves at a charge rate of 1 C are shown in Fig. 3.8. The peak at 19.15° corresponds to the LFP phases at 211 and 020, the peak at 19.5° to the FP phase at 211, and the peak at 19.85° to the FP phase at 020. A two-phase coexistence reaction thus appears to have occurred, given the decrease in the LFP phase peak and the increase in the FP phase peaks as the charging reaction proceeds. However, peak shifts were observed, especially at 60 nm. These represent a change in the lattice constants of the two phases, i.e., a change in the Li content. Next, we discuss the lattice constants and the half width at half maximum based on changes in peak location and peak width calculated from these measurement results.

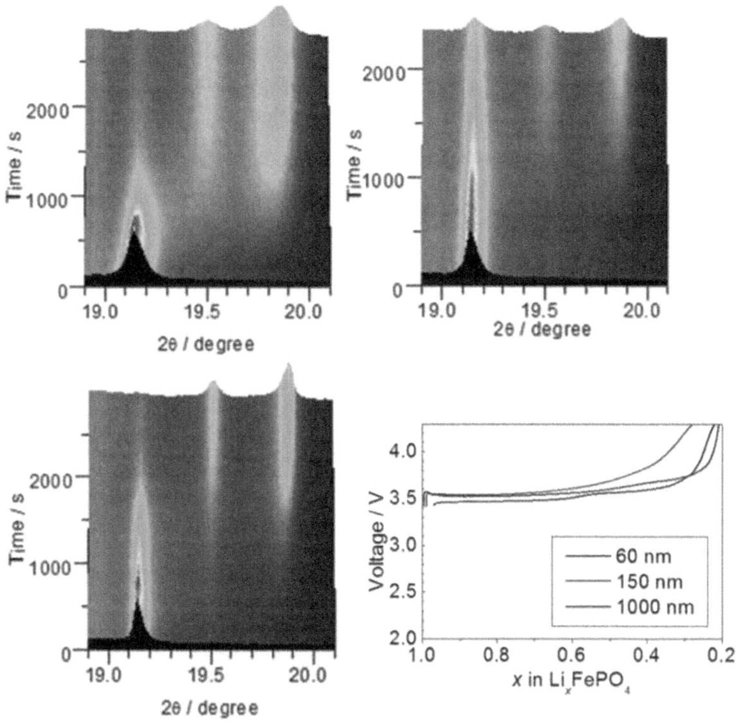

Figure 3.8 Time-resolved XRD patterns for $LiFePO_4$, for particle sizes of (a) 60 nm, (b) 150 nm, and (c) 1000 nm. (d) Charge profiles during time-resolved measurements.

Changes in the lattice constant during the 1 C charging reaction are shown in Fig. 3.9a. Charging pretreatment was performed in advance in these experiments, that is, charging until directly before the plateau region, in order to eliminate the influence of the solid-solution phase. For the LFP phase, lattice constant changes were seen at 60 nm. In contrast, at 150 nm and 1000 nm, lattice constant changes were not seen, confirming that variation is dependent on particle diameter. For the FP phase, on the other hand, the value that initially appears from the composition around $x = 0.9$ is larger than the lattice constant at equilibrium, and the value of the lattice constant visibly decreases as the reaction progresses. The change in the lattice constant in the initial period of FP phase formation is especially large. The same trend of a large decrease in lattice constant is visible at all particle diameters.

Half-width changes during the 1 C charging reaction are shown in Fig. 3.9b. The half width of the LFP phase can be seen to greatly increase, especially at a particle diameter of 60 nm. Reductions in the half width for the FP phase were observed for all samples. These trends are consistent with the changes in the lattice constant, with the half width seen to become smaller as the lattice constant approaches its equilibrium value. We believe these changes in the lattice constant reflect changes in the Li content in the two phases. Crystal distortion is also involved in half-width changes, but we believe that what was mainly observed were changes in crystal size.

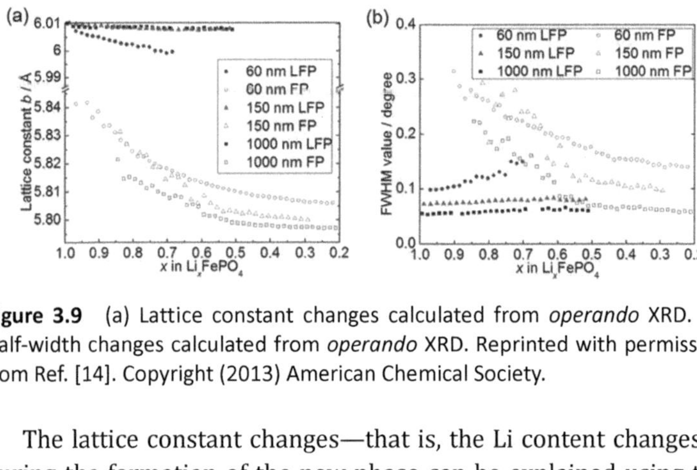

Figure 3.9 (a) Lattice constant changes calculated from *operando* XRD. (b) Half-width changes calculated from *operando* XRD. Reprinted with permission from Ref. [14]. Copyright (2013) American Chemical Society.

The lattice constant changes—that is, the Li content changes—during the formation of the new phase can be explained using the Gibbs energy diagram for the two-phase system, shown in Fig. 3.10. For the $LiFeO_4/FePO_4$ reaction system at equilibrium, a single-phase state is stable for compositions with $x = [1 - (1 - \alpha)]$ and with $x = (\beta - 0)$, and so the solid-solution reaction proceeds as follows: For compositions with $x = [(1 - \alpha) - \beta]$, the two-phase coexistence state is stable, and so the phases separate and a two-phase reaction occurs [13]. During the charge reaction (that is, the Li^+ extraction step), the phase transition begins at $x = (1 - \alpha)$ at equilibrium, and the reaction thereafter proceeds in the two-phase coexistence state with the LFP phase for $x = (1 - \alpha)$ and the FP phase for $x = \beta$. However, because nucleation of the FP phase in the real reaction is delayed and does not occur instantly, we believe that the phase transition may begin with $x = (1 - \alpha')$, beyond the stable composition. In this case, energy is gained to form the FP phase if the composition falls in the range from $x = \beta'$ (the point where the tangent line to the free energy curve for the LFP phase at $x = (1 - \alpha')$ intersects the free energy curve for the FP phase) to the stable composition at $x = \beta$. After FP phase formation, the compositions should then reach the thermodynamically stable composition of $x = \beta$. This phenomenon is also responsible for compositional changes in lithium being greatest at the time of new phase formation. In short, the system is delayed from reaching the stable composition because of deviations in the kinetic composition that occur in nonequilibrium states. These trends of compositional discrepancies in Li are consistent with empirical data; this is the first time such deviations in the kinetic composition of lithium have been observed in real operating cells.

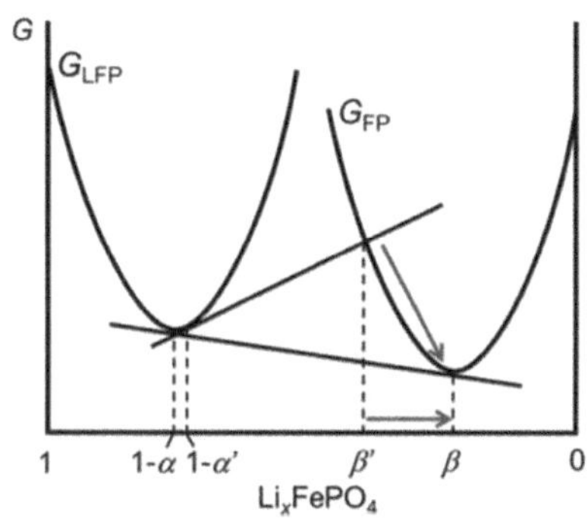

Figure 3.10 Relationships of Gibbs energy curves for $LiFePO_4$ and $FePO_4$ with phase changes. Reprinted with permission from Ref. [14]. Copyright (2013) American Chemical Society.

The above findings show that the FP phase first forms on LFP particles during this nonequilibrium phase transition. The FP phase in the initial stage of formation has a higher Li content than thermodynamically stable compositions. As charging proceeds, the Li content approaches the final, stable composition. The reaction proceeds independently on multiple particles. If charging were halted midreaction, then the phase boundary would remain present within individual particles. This would cause the interfacial energy to become unstable, resulting in relaxation of the phase boundary and its eventual disappearance. In this study, we observed changes in the lattice constant and the half-width value in the two-phase coexistence area for the first time. The lattice constant changes correspond to Li content changes between the two phases. The formation of the new, not-thermodynamically-most-stable phase can be explained in terms of kinetics. We thus observed, for the first time, phase transition behavior that is only visible through measurements at nonequilibrium and successfully constructed a dynamic model for the phase change.

We repeated five charge/discharge reaction cycles at a charge rate of 10 C; the time-resolved XRD profiles after 15 min. of relaxation are shown in Fig. 3.11. The peak at 19.15° corresponds to the LFP phases at 211 and 020, the peak at 19.5° to the FP phase at 211, and the peak at 19.85° to the FP phase at 020. A two-phase coexistence reaction appears to have occurred, given the decrease in the LFP phase peak and the increase in the FP phase peaks as the charging reaction proceeds. In addition, a new peak emerged near 19.35°, which originates from neither the LFP nor the FP phases. The new phase peak does not appear in the first charging step but grows in the discharge step and disappears in the charging steps that follow.

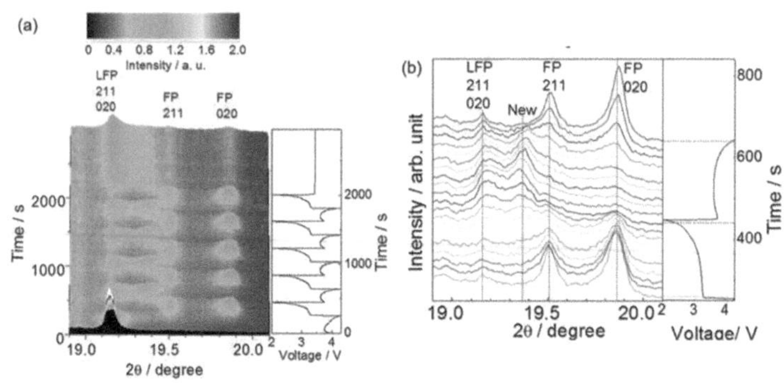

Figure 3.11 Time-resolved XRD patterns and charge/discharge curves following charge/discharge cycles at a rate of 10 C. Reprinted with permission from Ref. [15]. Copyright (2013) American Chemical Society.

However, it would be difficult to explain the cause of this new peak simply on the basis of a single diffraction line. Accordingly, we decided to perform time-resolved XRD measurements at other angular ranges at a charge rate of 10 C. From this, a similar new phase formation during discharge was also observed at diffraction lines of 200 and 301. We calculated the lattice constants of this new phase peak at 020, 200, and 301, assuming it was the same orthorhombic crystal as $LiFePO_4$ (i.e., a = 10.21 Å, b = 5.945 Å, and c = 4.750 Å). We found that the new phase has a lattice constant between that of the LFP phase and that of the FP phase. We hereby define this Li_xFePO_4 phase as the "L_xFP phase." The lattice constant of the L_xFP phase discovered in this study is extremely close to the value reported for the solid-solution phase of Li_xFePO_4, determined using high-temperature XRD (where x = 0.6–0.75) [17, 18]. Accordingly, the solid-solution phase of Li_xFePO_4 that appears stable at high temperatures (for x = 0.6–0.75) should likewise be formed during this nonequilibrium electrochemical reaction.

The formation of this L_xFP phase is dependent on speed. Figure 3.12 shows XRD patterns obtained immediately after the initial charge/discharge, during which most of the L_xFP phase is formed, for the same experiment at different charge/discharge rates. The L_xFP peak becomes clearly visible with higher charge rates. The L_xFP peak has completely disappeared from the XRD pattern that was made after relaxation of the same composition for over 24 h. Thus, the L_xFP phase is found to grow preferentially at high charge/discharge speeds, revealing a metastable phase that cannot be observed at equilibrium.

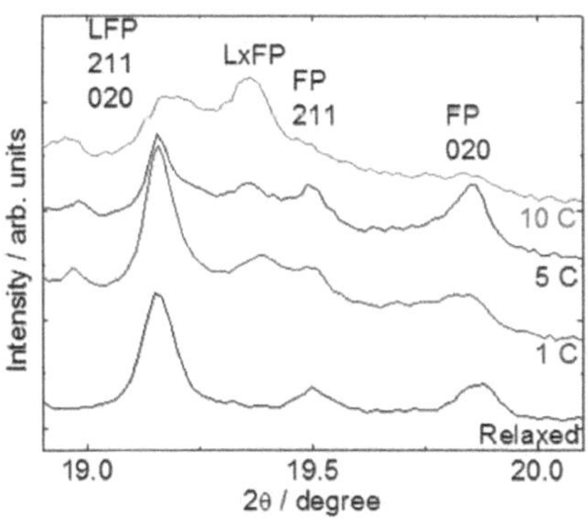

Figure 3.12 Charge/discharge speed dependence of L_xFP phase formation. Reprinted with permission from Ref. [15]. Copyright (2013) American Chemical Society.

The existence of the L_xFP phase is highly beneficial to the LFP/FP phase transition. The lattice constant gap between the LFP and FP phases is large—3.6% on the *b* axis and 1.8% on the *c* axis—increasing the interfacial energy associated with nucleation. Accordingly, a large current would not flow in a reaction where $FePO_4$ directly transitions to $LiFePO_4$. This makes the existence of the metastable Li_xFePO_4 phase quite significant. The lattice mismatch between Li_xFePO4 and $FePO_4$ is smaller than that between it and $LiFePO_4$—2.5% on the *b* axis and 0.66% on the *c* axis—making nucleation of the Li_xFePO4 phase more likely to occur. For the nucleation rate-limiting $LiFePO_4$, the reaction proceeds without issue, thanks to nucleation via the intermediate phase Li_xFePO_4. This is the reason that $LiFePO_4$ exhibits superior high-speed charge/discharge characteristics.

In this study, we discovered that Li_xFePO_4, a new phase having a lattice constant between those of $LiFePO_4$ and $FePO_4$, forms during high-speed charging/discharging of $LiFePO_4$. The amount of Li_xFePO_4 that appears is proportional to the reaction speed. The Li_xFePO_4 phase, which relaxes the lattice strain between the two phases, forms between them, and the reaction proceeds without issue. This intermediate phase appears to be what is primarily responsible for the high-speed charge/discharge behavior of $LiFePO_4$. In the time-resolved measurements of the phase transition behavior of $LiFePO_4$ in this study, we made direct observations of nonequilibrium phase behavior that heretofore could only be discussed in conjectural terms on the basis of steady-state measurements and calculations, and revealed heretofore-unknown phase transition phenomena. The results of our study are important to electrode development, providing clues for the construction of design guidelines for novel two-phase reaction-type electrodes.

3.4 Designing a Composite Electrode on the Basis of a Reaction Distribution Analysis

As the charge/discharge reaction proceeds, it causes the insertion and extraction of lithium ions within the electrode-active material. In practice, the electrode is a composite electrode, in which the active material is mixed three dimensionally to increase the loading rate. A composite electrode is a very complex structure. It is composed of an active material that stores lithium ions, a conductive material that forms electronic conduction paths, and a binding agent to join them, as well as an electrolyte that creates ionic conduction paths and exists in its vacant spaces. A schematic diagram of a composite electrode is shown in Fig. 3.13. It has been empirically recognized that the fabrication conditions of a composite electrode greatly affect its properties. Researchers have developed optimal electrode structures via trial and error in attempts to improve their performance.

However, to improve the characteristics of lithium-ion batteries to near their theoretical performance, we would need to clarify the factors governing these characteristics and then design a composite electrode that could generate the ideal reaction. From experiments on composite electrodes, we know that the reaction distribution has a deep cause-and-effect relationship on their performance. Complex electron and ion conduction paths are formed within composite electrodes. Nonuniformity in internal resistance within these paths is predicted to generate a reaction distribution—in particular, nonuniformity in ionic conduction resistance and electronic conduction resistance within the composite electrode [19–21]. These kinds of localized reactions due to the reaction distribution cause the utilization ratio of the electrode to fall, accelerating deterioration at specific sites. Therefore, understanding the conditions influencing the reaction distribution, and thus informing the design of composite electrodes with these data, is extremely important for their application. However, typical electrochemical measurement techniques gather voltage and current data for the whole composite electrode, so neither reaction sites inside the electrode nor nonuniformity in resistance can be observed directly.

In this study, we performed measurements of reaction distribution and separate internal resistances for a composite electrode and, by integrating and discussing the results, we aim to reveal the factors that govern the formation of a reaction distribution. Specifically, we assessed the electrochemical characteristics of several composite electrodes prepared at different electrode porosities and measured the reaction distributions in electrode cross sections using 2D-imaging XAS. Finally, we separately measured ionic conductivity and electronic conductivity inside the composite electrodes. The results revealed that the porosity of a composite electrode greatly affects the ionic conductivity within it, by acting to alter electrochemical characteristics as well as to form the reaction distribution in the electrode in the transverse direction.

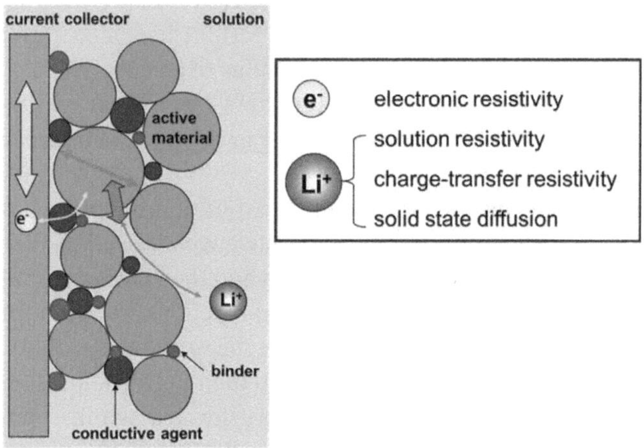

Figure 3.13 Schematic diagram of a composite electrode for a lithium-ion secondary battery and the electronic and ionic conduction paths within it.

Figure 3.14 shows 10 C discharge curves for $LiFePO_4$ electrodes with different porosities. The equilibrium potential of $LiFePO_4$ is ~3.45 V across most of the compositional range. The discharge voltage decreases as porosity increases, indicating an ohmic loss due to a drop in the degree of contact. Two factors may be responsible. The first is electronic resistance within the composite electrode. Contact between particles is not sufficient in composite electrodes with high porosities. This likely results in a failure to form good electronic conduction paths and a corresponding large rise in the effective electronic conduction resistance: hence, a large ohmic loss is produced. The other is contact resistance between the current collector and the composite electrode. In short, contact between the current collector and the composite electrode could have been insufficient, generating large contact resistance.

For electrodes with low porosities, a large voltage drop is observed at the end of the discharge and capacitance is found to decline. We surmise that this is because ionic conductivity within these composite electrodes is lower than within high-porosity composite electrodes. The electrolyte is retained in pores inside the composite electrode, but the pores in low-porosity electrodes are narrow and distorted. This makes it difficult for lithium ions to migrate within these narrow, distorted pores, potentially causing a large concentration gradient within the composite electrode. We believe that the large concentration polarization generated by such a concentration gradient causes a reaction distribution to arise and the rapid voltage drop at the end of the discharge to appear in the low-porosity electrodes. To clarify the above issues, we measured cross-sectional reaction distributions in electrodes by means of 2D XAS imaging.

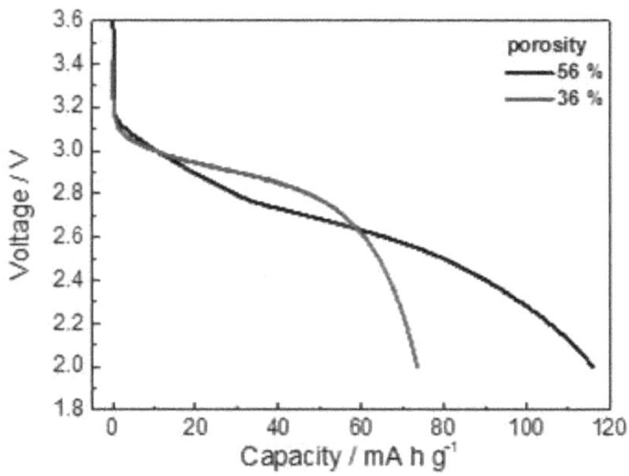

Figure 3.14 Discharge characteristics of $LiFePO_4$ cathodes of various porosities at a discharge rate of 10 C.

We used 2D XAS imaging to measure the cross-sectional reaction distribution in composite electrodes. The intensity of transmitted light was detected by means of a 2D sensor device with several X-ray detection elements in two dimensions. Incorporating the 2D sensor device allows the user to obtain local information, such as the electronic state(s) of specific elements at the location of interest and the ambient environment around those elements. A resolution of 10 × 10 μm was used in the empirical measurements to ensure data quality and precision [22]. Electrodes subjected to a charge/discharge treatment were processed using a microtome and placed in the apparatus. The electrode was irradiated by incident X-rays, while the 2D XAS spectra were measured by the 2D sensor device.

Composite electrodes with different porosities were discharged at a rate of 10 C until $Li_{0.5}FePO_4$ was obtained. The 2D XAS imaging results for cross sections at this time point are shown in Fig. 3.15. Absorption edge energy is mapped at every location, and the depth of discharge is visualized in the transverse direction. For composite electrodes with porosities of 36%, the reaction was found to proceed preferentially at the electrode/electrolyte interface; at porosities of 56%, the reaction proceeded uniformly. Thus, the 2D XAS imaging successfully visualized the generation of a significant reaction distribution with decreasing electrode porosity.

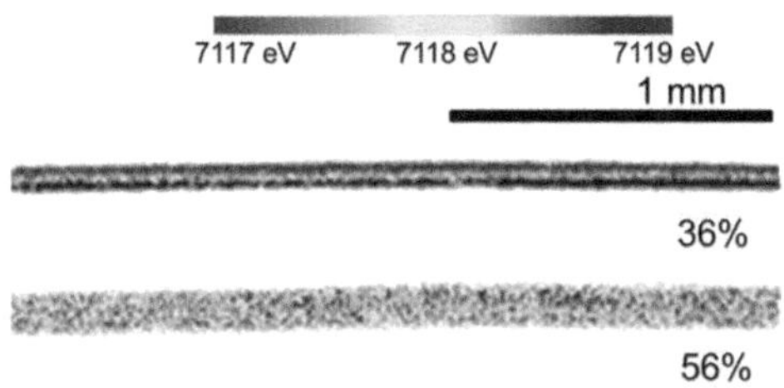

Figure 3.15 2D XAS imaging of cross sections of $LiFePO_4$ electrodes of various porosities (top: electrode surface side; bottom: collector foil side).

The reaction distribution is predicted to be determined by the balance of the electronic current and the ionic current within the composite electrode, and so the two must be examined separately. However, measuring electron conductivity separately from ion conductivity within a composite electrode is difficult because of the complex 3D structures involved.

Samples were prepared by coating an insulating substrate with sheet- or film-like electrode materials. Two potentiostats were wired to the system, one on the left and one on the right, and were operated after being set to the open-circuit voltage (OCV) value of the electrode fitted between the ionic conductor and electronic conductors. A difference arises in the electronic potential between the left and right working electrodes when a bias voltage is applied between them: the potential difference is the same as the bias voltage. Electronic current flows between the working electrodes according to this potential difference. The ionic and electronic potential differences are maintained at the OCV by the potentiostats, and so the difference in ionic potential between the left and right reference electrodes is the same as the bias voltage [23]. Ionic current flows between the counter electrodes, likewise, according to this potential difference. The potential profile has a linear shape, and so the ionic and electronic potential differences are maintained at the OCV throughout the sample. This setup prevents interconversion of the ionic current and the electronic current due to the electrochemical reaction.

Results for electronic conductivity and ionic conductivity are shown in Table 3.1. The effective electronic conductivity within the composite electrode increased linearly with decreasing porosity. It was around 100 mS/cm for the 60%-porosity electrode and 400 mS/cm for the 35%-porosity electrode. Electronic conductivity within the composite electrode was on the order of 100–400 mS/cm and increased with decreasing porosity. We believe this trend resulted from improved adhesion between the conductive aids, which reduced contact resistance. However, the electronic conductivity within the composite electrode was extremely high in this evaluation system, leading us to infer that it was not rate limiting.

The effective ionic conductivity within the composite electrode was ~5 mS/cm at porosities above 50%. In addition, ionic conductivity was ~2 mS/cm at 35% porosity. Ionic conductivities ranged from 1 to 6 mS/cm, with values 2 orders of magnitude smaller than the electronic conductivities. We presume this was because the apparent path for lithium-ion migration was longer (i.e., the migration difficulty was higher) for smaller pores, which reduced the empirically measured conductivity.

Table 3.1 Electronic conductivity and ionic conductivity for $LiFePO_4$ composite electrodes with different porosities

Porosity	35%	60%
Effective electronic conductivity	413(30) mS/cm	92(16) mS/cm
Effective ionic conductivity	1.9(0.9) mS/cm	5.4(2.3) mS/cm

We investigated how dependent the electronic and ionic conductivities within the composite electrode, as well as the rate characteristics, were on porosity. Ionic conductivity is rate limiting in the composite electrode; its value decreases with decreasing porosity. As shown in Fig. 3.16, the balance between electronic conduction resistance and ionic conduction resistance within the composite electrode varies greatly according to porosity. The porosity can cause a reaction distribution to form within the composite electrode, contributing to the dependence of the discharge capacity on porosity.

In summary, for low porosities, a large ionic potential distribution is generated within the electrode, as a result of which the reaction occurs at only the electrode/electrolyte interface and does not occur at all inside the active material. In this case, the discharge capacity is reduced. For high porosities, in contrast, the ionic potential distribution is small, which should make the reaction proceed uniformly and increase the available capacity. The differences in rate characteristics for various composite-electrode porosities have thus been demonstrated to be governed by ionic conductivity within the composite electrode.

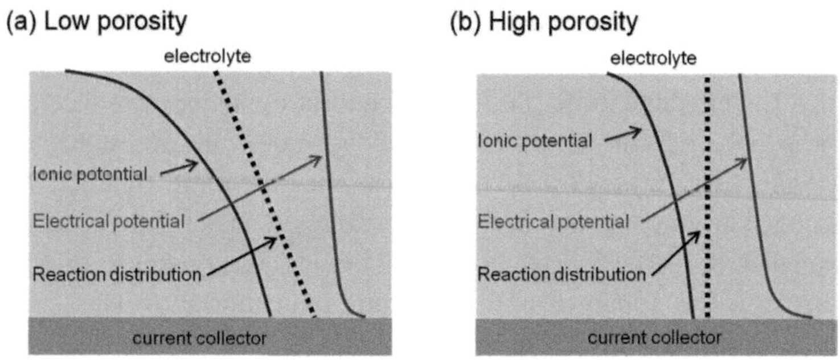

Figure 3.16 Illustration of ionic and electronic potential distributions and cross-sectional reaction distributions of a composite electrode.

3.5 Conclusion

Focusing on the hierarchical structure of lithium-ion secondary batteries, in this study we introduced a novel analysis method that possesses high temporal and spatial resolution. We also examined an applied case. Using the technique, we investigated the electrode/electrolyte interface, which serves as a reaction site; the phase change behavior of the active material; and reaction distributions in composite electrodes. We believe that high performance can be achieved on a rational basis, not through trial and error, by continuing with research and development after the reaction mechanism is clarified. These are valuable steps toward achieving novel, innovative lithium-ion secondary batteries with superior performance characteristics.

References

1. Armand, M. and Tarascon, J. M. (2008). Building better batteries, *Nature*, **451**, pp. 652–657.
2. Itagaki, M., Kobari, N., Yotsuda, S., Watanabe, K., Kinoshita, S. and Ue, M. (2005). $LiCoO_2$ electrode/electrolyte interface of Li-ion rechargeable batteries investigated by in situ electrochemical impedance spectroscopy, *J. Power Sources*, **148**, pp. 78–84.
3. Hirayama, M., Ido, H., Kim, K., Cho, W., Tamura, K., Mizuki, J. and Kanno, R. (2010). Dynamic structural changes at $LiMn_2O_4$/electrolyte interface during lithium battery reaction, *J. Am. Chem. Soc.*, **132**, pp. 15268–15276.
4. Hirayama, M., Yonemura, M., Suzuki, K., Torikai, N., Smith, H., Watkinsand, E., Majewski, J. and Kanno, R. (2010). Surface characterization of $LiFePO_4$ epitaxial thin films by X-ray/neutron reflectometry, *Electrochemistry*, **78**, pp. 413–415.
5. Matsui, M., Dokko, K. and Kanamura, K. (2008). Dynamic behavior of surface film on $LiCoO_2$ thin film electrode, *J. Power Sources*, **177**, pp. 184–193.
6. Wobrauschek, P. (2007). Total reflection x-ray fluorescence analysis: a review, *X-Ray Spectrom.*, **36**, pp. 289–300.
7. Takamatsu, D., Koyama, Y., Orikasa, Y., Mori, S., Nakatsutsumi, T., Hirano, T., Tanida, H., Arai, H., Uchimoto, Y. and Ogumi, Z. (2012). First in situ observation of the $LiCoO_2$ electrode/electrolyte interface by total-reflection X-ray absorption spectroscopy, *Angew. Chem. Int. Ed.*, **51**, pp. 11597–11601.
8. Yamamoto, K., Orikasa, Y., Takamatsu, D., Koyama, Y., Mori, S., Masese, T., Mori, T., Minato, T., Tanida, H., Uruga, T., Ogumi, Z. and Uchimoto, Y. (2014). Stabilization of the electronic structure at the cathode/electrolyte interface via MgO ultra-thin layer during lithium-ions insertion/extraction, *Electrochemistry*, **82**, pp. 897–900.
9. Orikasa, Y., Takamatsu, D., Yamamoto, K., Koyama, Y., Mori, S., Masese, T., Mori, T., Minato, T., Tanida, H., Uruga, T., Ogumi, Z. and Uchimoto, Y.

(2014). Origin of surface coating effect for MgO on $LiCoO_2$ to improve the interfacial reaction between electrode and electrolyte, *Adv. Mater. Interfaces*, **1**, p. 1400195.

10. Padhi, A. K., Nanjundaswamy, K. S. and Goodenough, J. B. (1997). Phospho-olivines as positive-electrode materials for rechargeable lithium batteries, *J. Electrochem. Soc.*, **144**, pp. 1188–1194.
11. Yamada, A., Chung, S. C. and Hinokuma, K. (2001). Optimized $LiFePO_4$ for lithium battery cathodes, *J. Electrochem. Soc.*, **148**, pp. A224–A229.
12. Chung, S. Y., Bloking, J. T. and Chiang, Y. M. (2002). Electronically conductive phospho-olivines as lithium storage electrodes, *Nat. Mater.*, **1**, pp. 123–128.
13. Yamada, A., Koizumi, H., Nishimura, S. I., Sonoyama, N., Kanno, R., Yonemura, M., Nakamura, T. and Kobayashi, Y. (2006). Room-temperature miscibility gap in Li_xFePO_4, *Nat. Mater.*, **5**, pp. 357–360.
14. Orikasa, Y., Maeda, T., Koyama, Y., Murayama, H., Fukuda, K., Tanida, H., Arai, H., Matsubara, E., Uchimoto, Y. and Ogumi, Z. (2013). Transient phase change in two phase reaction between $LiFePO_4$ and $FePO_4$ under battery operation, *Chem. Mater.*, **25**, pp. 1032–1039.
15. Orikasa, Y., Maeda, T., Koyama, Y., Murayama, H., Fukuda, K., Tanida, H., Arai, H., Matsubara, E., Uchimoto, Y. and Ogumi, Z. (2013). Direct observation of a metastable crystal phase of Li_xFePO_4 under electrochemical phase transition, *J. Am. Chem. Soc.*, **135**, pp. 5497–5500.
16. Orikasa, Y., Maeda, T., Koyama, Y., Minato, T., Murayama, H., Fukuda, K., Tanida, H., Arai, H., Matsubara, E., Uchimoto, Y. and Ogumi, Z. (2013). Phase transition analysis between $LiFePO_4$ and $FePO_4$ by in-situ time-resolved X-ray absorption and X-ray diffraction, *J. Electrochem. Soc.*, **160**, pp. A3061–A3065.
17. Chen, G. Y., Song, X. Y. and Richardson, T. J. (2007). Metastable solid-solution phases in the $LiFePO_4/FePO_4$ system, *J. Electrochem. Soc.*, **154**, pp. A627–A632.
18. Delacourt, C., Poizot, P., Tarascon, J. M. and Masquelier, C. (2005). The existence of a temperature-driven solid solution in Li_xFePO_4 for $0 < x < 1$, *Nat. Mater.*, **4**, pp. 254–260.
19. Doyle, M., Fuller, T. F. and Newman, J. (1993). Modeling of galvanostatic charge and discharge of the lithium/polymer/insertion cell, *J. Electrochem. Soc.*, **140**, pp. 1526–1533.
20. Ng, S.-H., La Mantia, F. and Novák, P. (2009). A multiple working electrode for electrochemical cells: a tool for current density distribution studies, *Angew. Chem. Int. Ed.*, **48**, pp. 528–532.

21. Hess, K. C., Epting, W. K. and Litster, S. (2011). Spatially resolved, in situ potential measurements through porous electrodes as applied to fuel cells, *Anal. Chem.*, **83**, pp. 9492–9498.

22. Katayama, M., Sumiwaka, K., Hayashi, K., Ozutsumi, K., Ohta, T. and Inada, Y. (2012). Development of a two-dimensional imaging system of X-ray absorption fine structure, *J. Synchrotron Radiat.*, **19**, pp. 717–721.

23. Siroma, Z., Hagiwara, J., Yasuda, K., Inaba, M. and Tasaka, A. (2010). Simultaneous measurement of the effective ionic conductivity and effective electronic conductivity in a porous electrode film impregnated with electrolyte, *J. Electroanal. Chem.*, **648**, pp. 92–97.

Chapter 4

Introduction to Ion Beam Analysis

Tomihiro Kamiya,[a] **Takahiro Satoh,**[b] **and Akiyoshi Yamazaki**[c]
[a]*Graduate School of Science and Technology, Gunma University, 1-5-1 Tenjin-cho, Kiryu, Gunma Prefecture, 376-8515 Japan*
[b]*Takasaki Advanced Radiation Research Institute, Quantum Beam Science Research Directorate, National Institutes for Quantum and Radiological Science and Technology, 1233 Watanuki-machi, Takasaki, Gunma Prefecture, 370-1292 Japan*
[c]*Faculty of Pure and Applied Sciences, University of Tsukuba, 1-1-1 Tennodai, Tsukuba, Ibaraki Prefecture, 305-8577 Japan*
kamiya.tomihiro@gunma-u.ac.jp

4.1 Ion Beam Analysis

4.1.1 Introduction

Ion beam analysis techniques were developed and have been improving since the 1960s, together with improvement in radiation detectors [1]. Since mega-electron-volt (MeV) energy ion beams induce various kinds of interactions with matter, many analysis methods are performed, such as particle-induced X-ray emission

Lithium-Ion Batteries: Overview, Simulation, and Diagnostics
Edited by Yoshiaki Kato, Zenpachi Ogumi, and José Manuel Perlado Martin

ISBN 978-981-4800-40-2 (Hardback), 978-0-429-25934-0 (eBook)
www.panstanford.com

(PIXE) analysis, particle-induced γ-ray emission (PIGE) analysis, nuclear reaction analysis (NRA), Rutherford backscattering spectroscopy (RBS), and secondary ion mass spectroscopy (SIMS).

Due to the long range of MeV ion beams, they induce intense energy transfer in a region with a high aspect ratio along the long ion trajectory. Therefore, relatively thick samples can be analyzed without elaborate preparation processes, such as thinning or homogenizing. For example, Sera et al. developed a technique with PIXE for quantitative analysis of human hair without any prior process, revealing dependencies of the trace element concentrations on the living environment and daily foods [2]. It is also possible to determine the chemical state and elemental concentrations in biological cells and in aerosol particles under the environment close to the actual by measuring ion luminescence and PIXE at the same time [3]. Furthermore it is possible to obtain multielemental distributions with high spatial resolution by using ion microbeams [4, 5]. 3D analyses of minute samples like biological cells and aerosols are also possible [6, 7].

When an ion is incident in matter, it loses its energy by interaction with atoms. The energy loss per distance $-(dE/dx)$ is called stopping power. There are two main interactions causing energy loss of an ion beam in matter. One is elastic scattering, which is dominant for low-energy particles. Another is inelastic scattering, which is dominant for high-energy particles. The main interaction for the inelastic scattering is the ionization interaction, where the charged particle excites or ionizes the bound electrons of the atoms through coulomb interaction.

When the ion energy is very high, where the velocity of the projectile ion is much higher than the velocity of the bound electrons, the stopping power $-(dE/dx)$ is given by the following formula derived by Bethe [8]

$$-\frac{dE}{dx}=\frac{4\pi}{m_e c^2}\frac{n_t Z_p^2}{\beta^2}\left(\frac{e^2}{4\pi\varepsilon_0^2}\right)^2\left[\ln\frac{2m_e c^2\beta^2}{I\left(1-\beta^2\right)}-\beta^2\right] \tag{4.1}$$

where the ion with speed v_p and charge $Z_p e$ travels in a material with an electron density n_t and a mean excitation potential I. c is the speed of light, $\beta = v_p/c$, and ε_0 is the permittivity of vacuum. Here n_t is given by

$$n_t = \frac{N_A Z_t \rho}{A_t M_u}$$

with ρ the density of the material, Z_t its atomic number, A_t its relative atomic mass, N_A the Avogadro number, and M_u the molar mass constant. The mean excitation potential I (in eV) is approximately given by

$$I = 10Z_t. \qquad (4.2)$$

This equation is valid for the ions with energy over ~400 keV/nucleon. For the MeV-energy ions, Eq. 4.1 shows that the stopping power is proportional to the square of the particle charge, but it does not depend on the particle mass. It decreases with the particle velocity v_p as $1/v_p$, which means that it is inversely proportional to the particle energy. The range of the ion, which gives the traveling distance of the ion in materials, is derived from the stopping power by

$$R = \int_0^E \frac{1}{dE/dx} dE$$

Since Eq. 4.1 is not valid for projectile ions with lower energy (less than ~100 keV/nucleon) and also for heavy ions with a high Z_p, several corrections have been made to this formula in order to improve the agreement with experimental data [9–11]. A well-known simulation program TRIM/SRIM (acronym for TRansport of Ions in Matter and the more recent version Stopping and Range of Ions in Matter) has been utilized to reproduce propagation of ions in matter [12].

4.1.2 Analyses Based on Sputtering and Elastic Ion Scattering

Low-energy ion beams of around 10 keV/nucleon interact with the surface atoms, leading to sputtering due to collision with the surface atoms followed by their being flicked out of the surface as secondary ions. These secondary ions can be ionized atoms or atomic and molecular clusters. Elemental or chemical states of the specimen surface are analyzed by SIMS, where the masses of these secondary ions are analyzed with electromagnetic or time-of-flight mass spectrometers. Since the detection efficiency depends on the species, energy, and incidence angle of the incident ions, optimum conditions have to be selected for the target sample to be analyzed.

Ion beams of MeVs/nucleon mainly induce ionization by colliding with electrons in the target. The incident ions interact also with the nucleus, causing elastic and inelastic scatterings. The measurement of the elastic backscattered ions is called RBS. In RBS, the mass of the target atoms is determined from the energy of the backscattered ions.

In the elastic scattering, the momentum and energy of the incident ion and the target atom are preserved. When the incident ion with mass m_1 and energy E_0 collides with the target atom with mass m_2 and scattered to the direction θ_1 in the laboratory frame, as shown in Fig. 4.1, the energy of the scattered ion E_1 is described by the following equation:

$$E_1 = \left(\frac{m_1 \cos\theta_1 + \sqrt{m_2^2 - m_1^2 (\sin\theta_1)^2}}{m_1 + m_2} \right)^2 E_0 \equiv kE_0 \tag{4.3}$$

where k is the kinematic factor and $m_1 < m_2$ is assumed. Therefore, the mass of the target atom m_2 can be determined by measuring E_1 and θ_1. The differential cross section due to coulomb scattering is described by the Rutherford scattering formula

$$\frac{d\sigma}{d\Omega} = \left(\frac{Z_1 Z_2 e^2}{4E_0} \right)^2 \frac{1}{\left(\sin\theta/2\right)^4} \tag{4.4}$$

where Z_1 and Z_2 are the atomic numbers of the incident and the target atoms, respectively, and θ is the scattering angle in the center of the mass frame. The scattering cross section is proportional to the square of Z_1 and Z_2 and inversely proportional to the square of the incident energy E_0. Figure 4.2 is a schematic RBS spectrum in the case of a thin target with a simple substance such as a silicon wafer, where E_d is the energy of the scattered particle from the deepest side, as shown in Fig. 4.3. The thickness x of the target is described as follows:

$$x = (kE_0 - E_d) / \left[\frac{kS(E_0)}{\cos\theta_2} + \frac{S(E_d)}{\cos\theta_3} \right] \tag{4.5}$$

where $S(E_d)$, θ_2, and θ_3 are the stopping power at the deepest side, the incidence angle, and the scattering angle, respectively, as shown in Fig. 4.3.

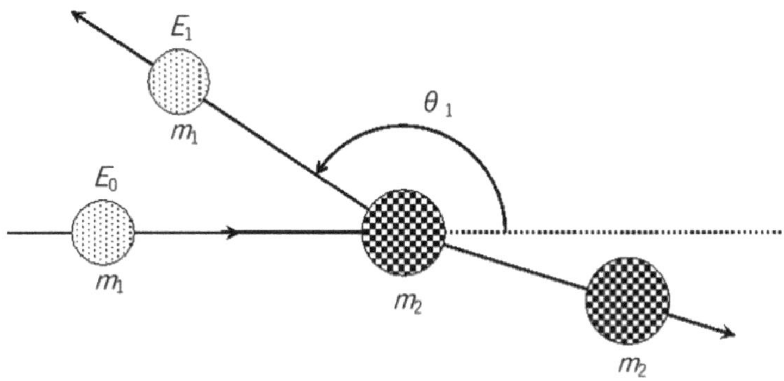

Figure 4.1 Schematic diagram of the elastic scattering, shown in the laboratory frame, between an incident ion with mass m_1 and a target atom with mass m_2.

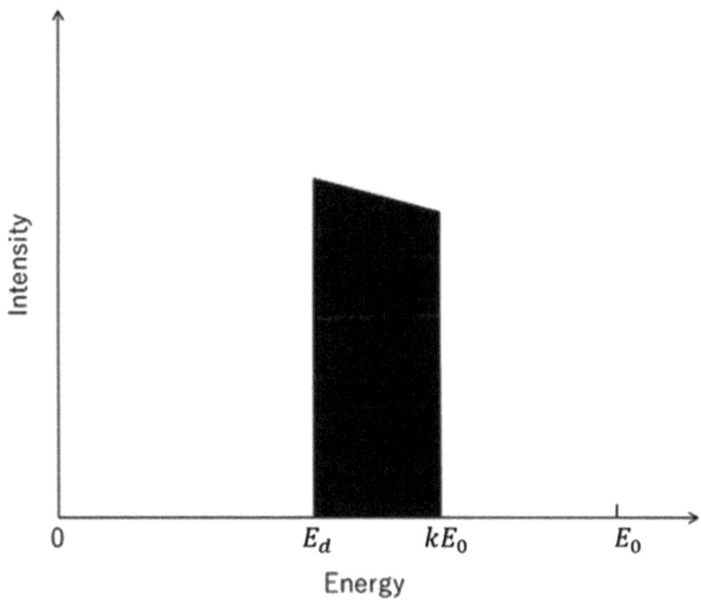

Figure 4.2 Schematic RBS spectrum in the case of a thin target composed of a simple substance.

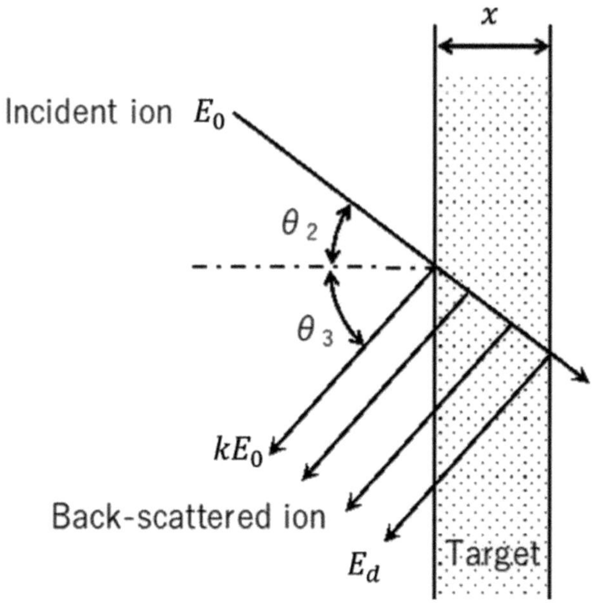

Figure 4.3 Schematic diagram of the elastic scattering of the incident ion by a thin solid target.

In the case of a thick target, the energy loss of the incident ion along the pass from the surface to the target atom and the energy loss of the scattered ion from the target atom to the surface have to be calculated in data analysis. The RBS spectrum becomes quite complex if the target has multiple layers of compounds. Therefore, spectral fitting using analysis software such as simulation for nuclear reaction analysis (SIMNRA) [13] is necessary for actual depth analysis. Data analysis by SIMNRA is described in Section 5.2.5 (Chapter 5).

For RBS analysis, it is desirable that the target elements are distributed homogeneously in the host material, such as a silicon crystal homogeneously bombarded by an ion beam. The RBS technique can be utilized also to investigate the location of the impurity element in the crystal lattice. When an ion beam with small divergence is incident on a single crystal, the range becomes longer when the incidence angle of the beam is aligned to the direction of the crystal lattice. This effect, called "channeling," is useful, for example, for analyzing impurities and to study defects in the crystal.

4.1.3 Analyses Based on Ionization Interaction

The MeV-energy ion beam has a large cross section for exciting the inner shell electrons of a target atom in material. Material analysis by measuring characteristic X-rays emitted from the inner shell electrons is called PIXE. PIXE analysis is utilized for highly sensitive trace element analysis in medicine, pharmacy, biology, and environmental science. PIXE analysis was carried out first in 1970 by Johansson et al. [14]. Since it was shown that only 10^{-12} g of titanium and calcium was detectable with this method, research and development on PIXE was accelerated as a new analysis method [15].

A schematic diagram of the generation of characteristic X-rays is shown in Fig. 4.4. When a light ion interacts with an atom, an electron is ejected from the inner shell. The outer-shell electron transits to the inner shell, emitting X-rays having an energy equivalent to the energy difference between the inner and outer shells. Since each element has a unique set of energy levels, the energy of the emitted X-rays is characteristic to each element. The typical transitions to K, L, and M shells emitting K-, L-, and M-X-ray lines, respectively, are shown in Fig. 4.5.

Figure 4.6 shows an example of the PIXE spectrum of aerosol particles collected on a membrane filter, which was irradiated with a 3 MeV proton beam extracted through a thin window into the atmosphere containing argon. The intensities of many peaks in the PIXE spectrum reveal the elemental concentrations in the sample, such as Ca, Ti, Mn, Fe, Cu, Zn, and Br. In addition, there are two "escape peaks," at 1.2 keV and 1.5 keV, which often appear when an external beam and Si(Li) detectors are used. When an incoming *E* keV X-ray ionizes the K shell of a Si atom in the detector, Si K_α X-ray of 1.74 keV escapes and *E*.– 1.74 keV energy is deposited in the detector. The two escape peaks in Fig. 4.6 were actually induced by the Ar K_α (2.96 keV) and K_β (3.19 keV) X-rays.

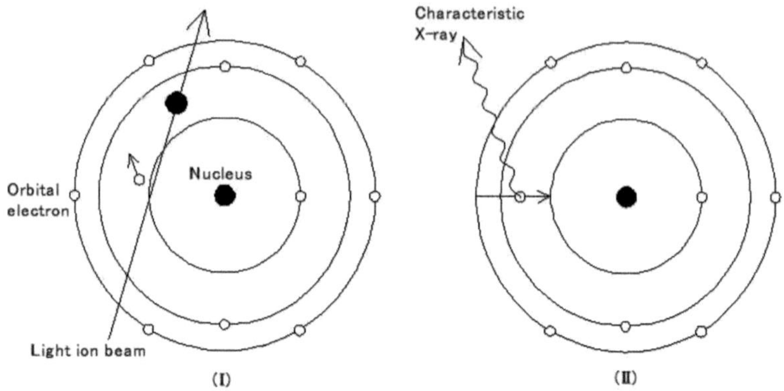

Figure 4.4 (I) Inner shell ionization on impact of a light-ion beam, followed by (II) emission of characteristic X-rays by electron transition.

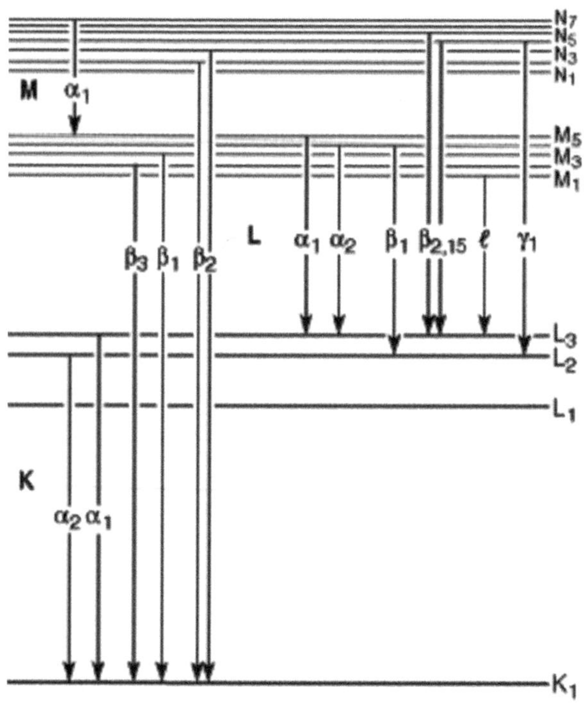

Figure 4.5 Transitions that give rise to the characteristic emission lines [16]. When a vacancy is created in the K shell and filled by an electron from the L or M shells, K_α or K_β X-rays are emitted.

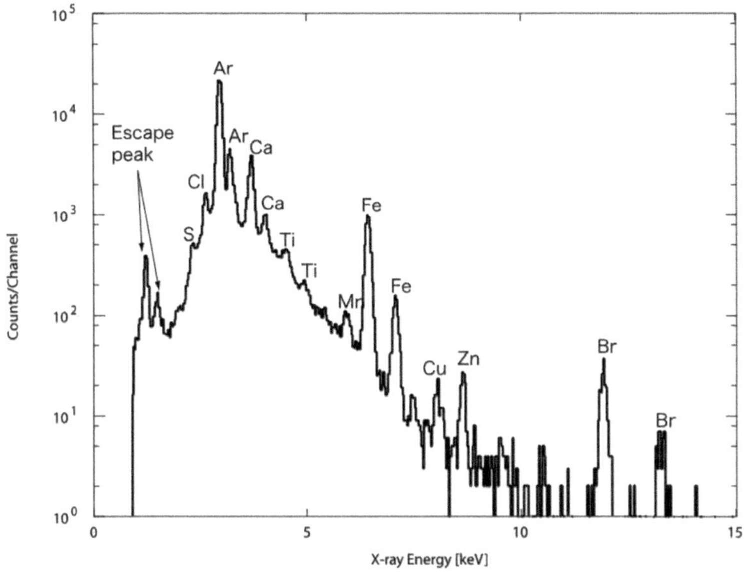

Figure 4.6 The PIXE spectrum of an aerosol sample generated by an external 3 MeV proton beam. The membrane contained bromine (Br). Two argon peaks appeared due to 0.93% by volume of argon (Ar) in the atmosphere.

Figure 4.7 shows a schematic diagram of a PIXE analyzer. The ion beam accelerated by an electrostatic accelerator passes through an analyzing magnet for energy selection and is guided to a target. Downstream of the target, the beam is received by a Faraday cup for measuring the beam current. Since the sensitivity of PIXE analysis increases between 2 and 4 MeV/nucleon, proton accelerators up to 3 MeV are often used for PIXE analysis. The X-ray emitted from the sample is detected by a solid-sate detector, and its energy is converted into a voltage signal and digitized by a multichannel analyzer to provide the X-ray spectrum. A Si(Li) semiconductor detector is often used as an X-ray detector. The X-rays are incident to the detector, located in vacuum, through a thin window of beryllium or polymer. The detection efficiency (eff) for the X-ray energy ($\hbar\omega$) is given by

$$\mathrm{eff}(\hbar\omega) = e^{-(\mu_{\mathrm{Be}}x_{\mathrm{Be}} + \mu_{\mathrm{Au}}x_{\mathrm{Au}} + \mu_{\mathrm{Si}}x_{\mathrm{Si}})}(1 - e^{-\mu_{\mathrm{Si}}x_{\mathrm{D}}}) \tag{4.6}$$

where μ_{Be}, μ_{Au}, and μ_{Si} are X-ray linear absorption coefficients for beryllium, gold, and silicon and x_{Be}, x_{Au}, x_{Si}, and x_D are thicknesses of the beryllium window, the thin gold electrode, and the dead and the depletion regions in the silicon crystal, respectively.

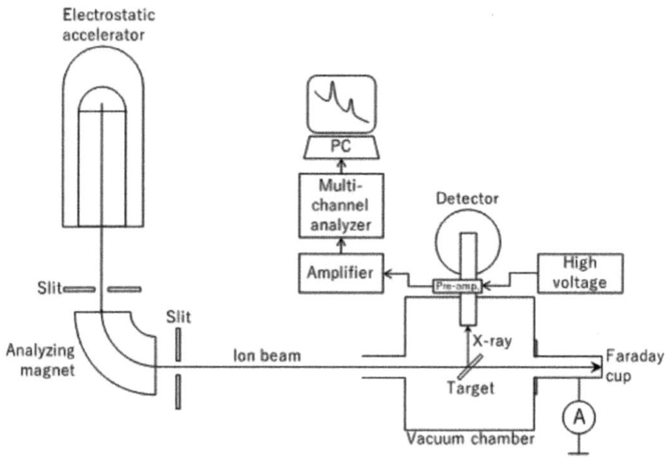

Figure 4.7 Schematic diagram of a PIXE analyzer.

Figure 4.8 shows the efficiency curves for x_{Be} = 8, 13, and 25 μm and x_D = 1, 3, and 5 mm. Although PIXE is not suitable to measure the lithium concentration since the 0.054 keV K_α X-ray of lithium is absorbed in the window, PIXE is suitable to measure concentrations of various elements, such as manganese, iron and cobalt contained in the positive electrode of a Li-ion battery, with a high sensitivity, at the parts per million level .

A reference method, using a sample containing a standard element of known amount, is popular for quantitative analyses by PIXE. The count ratio (Y_A/Y_S) of element A to the standard element S is given by the following equation:

$$\frac{Y_A}{Y_S} = \frac{M_S m_A \mathrm{eff}(\hbar\omega_A)\sigma_A(E_P)Q_A}{M_A m_S \mathrm{eff}(\hbar\omega_S)\sigma_S(E_P)Q_S}, \quad 4.7)$$

where M, m, $\sigma(E_p)$, and Q are atomic mass, element content, total production cross section of characteristic X-ray at particle energy E_p, and amount of irradiation, respectively. In the internal standard method where the standard element S is uniformly mixed in a target in advance, Q_A and Q_S are equal. For the K_α and K_β X-rays, the production cross sections (σ_{K_α} and σ_{K_β}) are given, respectively, by:

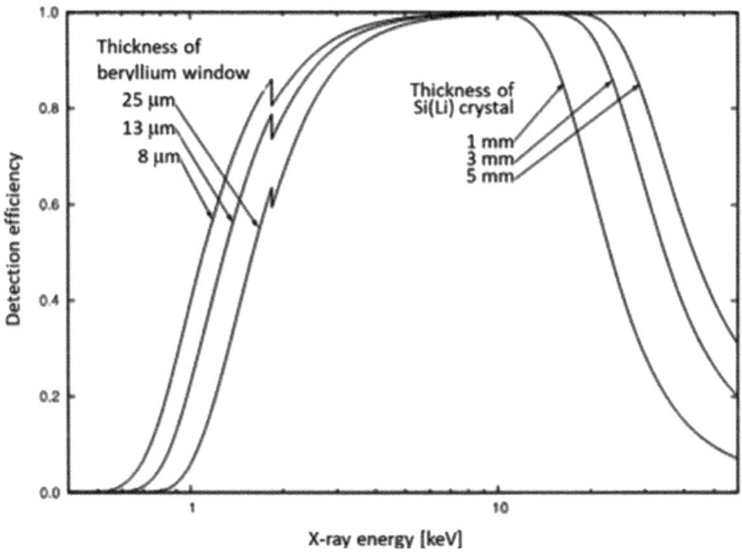

Figure 4.8 Detection efficiency of Si(Li) detectors for various Be filter thicknesses and depletion region thicknesses of the Si detector.

$$\sigma_{K_\alpha}(E_P) = \sigma_i(E_P)\omega_K\left(\frac{K_\alpha}{K_\alpha + K_\beta}\right) \tag{4.8}$$

and

$$\sigma_{K_\beta}(E_P) = \sigma_i(E_P)\omega_K\left(\frac{K_\beta}{K_\alpha + K_\beta}\right) \tag{4.9}$$

where σ_i, ω_K, and $K_\alpha : K_\beta$ are K-shell ionization cross section, fluorescence yield, and intensity ratio of the K_α and K_β X-rays, respectively. σ_i can be computed by a program based on energy-loss coulomb-repulsion perturbed-stationary-state relativistic (ECPSSR) theory [17–19]. ω_K and $K_\alpha : K_\beta$ are given by a program based on databases of physical data [20]. The detection limit in PIXE analysis is determined by a continuous background underlying the peaks of the characteristic X-rays. The peaks in an X-ray spectrum are discriminable if the net count of the characteristic X-rays (Y_A) is three times larger than the statistical error of the background ($\sqrt{Y_B}$), represented by the following equations:

$$Y_A \geq 3\sqrt{Y_B} \ (Y_B > 0), \tag{4.10}$$

$$Y_A \geq 3 \ (Y_B > 0). \tag{4.11}$$

Figure 4.9 represents the detection limit for the specific elements of $Z = 10$–90 in the case of proton bombardment with an energy of E_p = 0.5–5 MeV to biological samples [21].

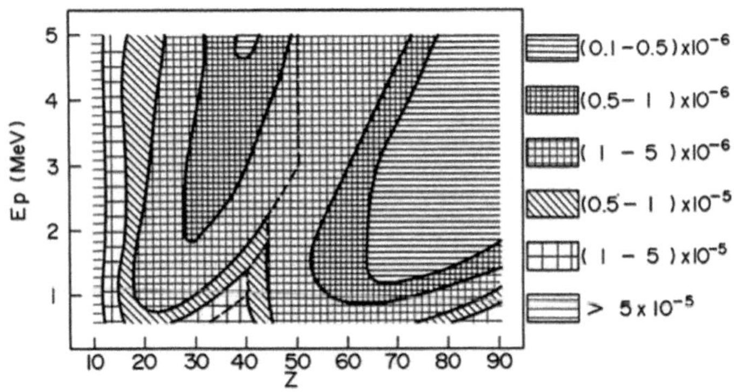

Figure 4.9 Detection limit for PIXE analysis of biological samples by proton bombardment. Reprinted from Ref. [21], Copyright (1988), with permission from Elsevier.

4.1.4 Analyses Based on Nuclear Excitation and Nuclear Reaction

Inelastic particle interaction can also take place in nuclear excitations and nuclear reactions. It is possible to detect trace elements in samples by observing the γ-rays emitted during relaxation of the nuclear excitation or by observing fragment particles or γ-ray radiations emitted from the radio isotopes generated by a nuclear reaction. Those are utilized as PIGE for analysis of light elements, such as fluorine and lithium. These light elements are difficult to detect with PIXE analysis since the low-energy-characteristic X-rays are absorbed in the sample and a beryllium window of the detector.

PIGE analysis is a complementary method to PIXE analysis since it can detect low-atomic-number elements, which cannot be detected by PIXE analysis. The depth information of the element can be obtained when the reaction takes place under the resonance condition at the particular colliding energy. Nuclear excitation takes place in this case when the incident particle reaches the resonance energy, after gradually losing energy during traveling in the material. When the γ-ray due to the resonant interaction is measured, the depth of the detected element is confined to a thin layer with a thickness that depends on the resonance energy width. The measured counts of the γ-rays are basically proportional to the elemental concentration in the target, though the obtained spectra are not so straightforward as in PIXE and RBS analysis.

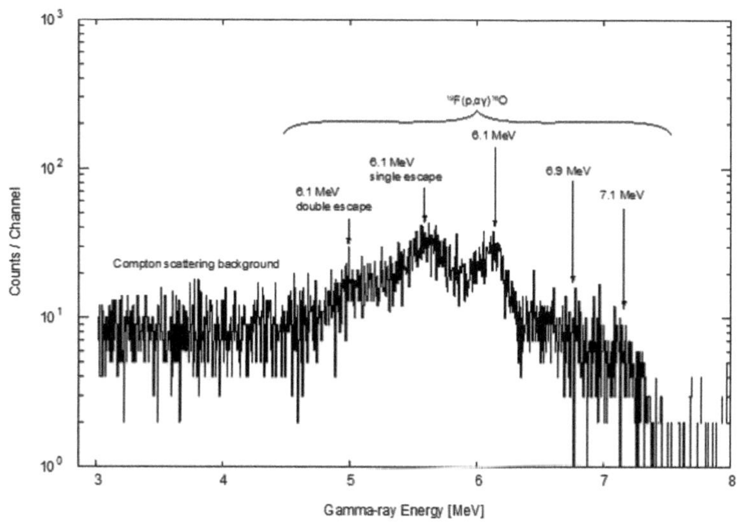

Figure 4.10 A PIGE spectrum of a tooth sample generated by a 1.7 MeV proton beam and measured by a NaI(Tl) scintillation detector. The 6.1 MeV, 6.9 MeV, and 7.1 MeV γ-rays of $^{19}F(p,\alpha\gamma)^{16}O$ in fluoride painting are detected.

In PIGE analysis, high-purity germanium (HPGe) detectors or thallium-activated sodium iodide—NaI(Tl)—scintillation detectors are generally used for detecting γ-rays, because it is difficult to make silicon detectors thicker than a few millimeters. The HPGe detectors have high energy resolution and relatively high detection efficiency. They are most suitable for γ-rays with energy of less than 1 MeV such as the 427 keV γ-ray from $^{7}Li(p,p'\gamma)^{7}Li$ reaction. On the other hand, although the energy resolution is lower than that of HPGe detectors, NaI(Tl) detectors are often used because of their higher detection efficiency for γ-rays of a few MeVs by using thick NaI(Tl) crystals. Figure 4.10 displays an example of a PIGE spectrum, obtained with a 4 in. × 4 in. NaI(Tl) detector, of a tooth coated with fluoride paint.

The γ-rays relevant to elemental analysis are given in Table 4.1, where the corresponding nuclear reactions are shown. The cross-section data for proton-induced γ-ray emission could be found in the nuclear data services of the International Atomic Energy Agency (IAEA, https://www-nds.iaea.org/). Moreover, the reference database for PIGE has been developed in a coordinated research program of the IAEA (https://www-nds.iaea.org/pige/).

Table 4.1 Proton-induced reactions useful for elemental analysis of light elements ($Z \leq 9$)

Element	γ-ray energy (keV)	Reaction	Note
Li	429	$^{7}Li(p,n\gamma)^{7}Be$	$E_p > 3$ (MeV)
	478	$^{7}Li(p,p'\gamma)^{7}Li$	
Be	3562	$^{9}Be(p,\alpha\gamma)^{6}Li$	
B	429	$^{10}B(p,\alpha\gamma)^{7}Be$	
	718	$^{10}B(p,p'\gamma)^{10}B$	
	2125	$^{11}B(p,p'\gamma)^{11}B$	$E_p > 3$ (MeV)
C	3089	$^{13}C(p,p'\gamma)^{13}C$	$E_p > 3.5$ (MeV)
	4439	$^{12}C(p,p'\gamma)^{12}C$	$E_p > 5.4$ (MeV)
N	2321	$^{14}N(p,p'\gamma)^{14}N$	
	4439	$^{15}N(p,\alpha\gamma)^{12}C$	
O	495	$^{16}O(p,\gamma)^{17}F$	
	871	$^{17}O(p,p'\gamma)^{17}O$	
	1982	$^{18}O(p,p'\gamma)^{18}O$	
	6129	$^{16}O(p,p'\gamma)^{16}O$	$E_p > 7.5$ (MeV)
F	110	$^{19}F(p,p'\gamma)^{19}F$	
	197	$^{19}F(p,p'\gamma)^{19}F$	
	6129	$^{19}F(p,\alpha\gamma)^{16}O$	

A typical experimental setup for detecting secondary particles including reaction products for NRA is schematically shown in Fig. 4.11. An ion beam of >1 MeV energy bombards a sample normally, and the energy spectra of the charged particles due to Rutherford backscattering or generated as nuclear reaction products are measured. The charged particle detectors cover a solid angle Ω at a certain detection angle θ to the beam axis. Although a higher intense signal output is obtained with a larger Ω, the energy resolution is reduced because the particle energy depends on the emission angle as described in Ref. [22]. On the other hand, when we choose a smaller Ω, the spectral resolution becomes higher.

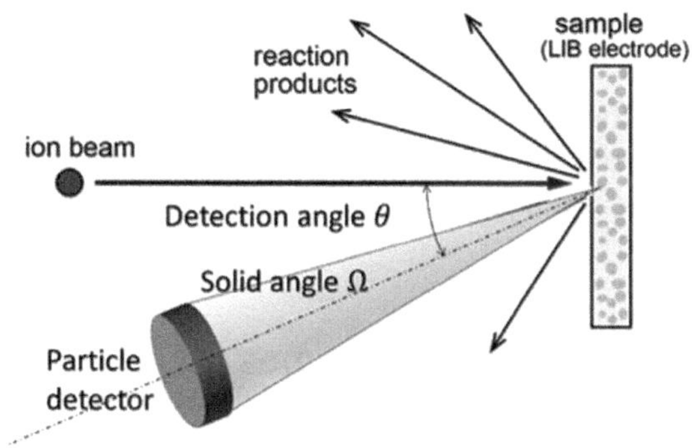

Figure 4.11 A schematic layout for nuclear reaction analysis by detecting secondary particles including reaction products.

The particle energy spectrum consists of two groups. One is the backscattered projectile (beam ion), which is used for RBS diagnostics. The other is the nuclear reaction products, which are used for the NRA. In NRA, applicable nuclear reactions for elemental analysis are classified into two groups.

One is the resonant nuclear reaction, in which the excitation function (dependence of the reaction probability on the energy of the projectile) has several sharp peaks, which are called "resonance." One of the well-known examples of the resonant NRA (RNRA) is $^{1}H(^{15}N,\alpha\gamma)^{12}C$, which is often used for hydrogen depth profiling [23, 24]. The excitation function of this reaction has an extremely sharp peak at 6.385 MeV of the incident 15-nitrogen energy. The width of the resonance is only 1.8 keV [25].

This method has been widely applied to characterize hydrogen contamination on metal surfaces [26–29]. A detailed description of the $^{1}H(^{15}N,\alpha\gamma)^{12}C$ nuclear reaction can be found in a review article by Lanford [30] and papers cited therein. In addition, various types of NRA for hydrogen detection are described in the review article by Ziegler et al. [31]. The other is the nonresonant nuclear reaction, in which the excitation function does not have a sharp peak. When nonresonant nuclear reactions are used, depth profiling can be done by observing the energies of the reaction products. Examples for measuring the depth profiles of light elements are as follows: $^{7}Li(p,\alpha)^{4}He$, $^{9}Be(p,\alpha)^{6}Li$, $^{11}B(p,\alpha_1)^{8}Be$, $^{19}F(p,\alpha)^{16}O$, $^{12}C(d,p)^{13}C$, $^{16}O(d,p_1)^{17}O$, $D(^{3}He,\alpha)^{1}H$, $^{11}B(^{3}He,p)^{13}C$, $^{16}O(^{3}He,\alpha)^{15}O$, and $^{31}P(^{4}He,p)^{34}S$.

To obtain a depth profile, it is necessary to know the excitation function and the stopping power of the reaction products in the sample. Details of the nonresonant NRA depth profiling applied to the Li-ion battery diagnostics are described in Section 5.2 (Chapter 5).

4.2 Accelerator Technologies for Ion Beam Analysis

4.2.1 Introduction

An ion with mass m and electric charge q is accelerated in an electric field $\mathbf{E}$ as

$$m\frac{d^2\boldsymbol{r}}{dt^2}=q\mathbf{E} \tag{4.12}$$

where r is the position of the ion. An ion beam is a flow of the charged particles collectively accelerated with an accelerator. The ion accelerator consists of an ion source, an accelerator with a high-voltage generator, a beam transport system, and end stations, which are combined in vacuum. The ion beam is characterized by the ion mass depending on the element (atom or cluster), the charge, the energy of individual particles, the intensity (or current), and the beam size.

Accelerators are classified into two types: electrostatic accelerators [32, 33] and oscillating field accelerators [34, 35]. In the electrostatic accelerator, the electric field **E** is generated by applying a constant high-voltage V to the electrode. The maximum voltage is limited to ~10 MeV due to electrical breakdown. In the oscillating field accelerator, where a radio frequency (RF) electric field is applied to the ions, it is possible to circumvent the breakdown and accelerate the ions to more than several tens of GeV to be used for high-energy physics research.

The MeV-energy ion beams have a broad range of applications in material science, such as material analyses and radiation effects, fabrication and processing of microelectronics, and biomedical research. The low-energy ion beams, of less than 100 keV, generated by smaller-scale electrostatic accelerators, are widely used in material science and in industry, especially in semiconductor processing.

Many accelerators that generate ion beams of approximately MeVs/nucleon energy have been built for nuclear physics after Cockcroft and Walton succeeded in artificially changing lithium nucleus to other nuclei by using accelerated proton beam in 1932 [33]. Since then, MeV-energy ion beams are finding applications in broader fields: analyses of materials and biomedical samples, material processing, radioisotope production for medical imaging, and testing of radiation effects on materials, electric devices for space or nuclear energy development, and biological systems for medical therapy.

The trajectories of the charged particles can be precisely manipulated by electromagnetic fields. The ion microbeam technology is one of the ultimate techniques to control the ion beam. Ion microbeam facilities are located all over the world (as shown in Fig. 4.12), including Japan (Takasaki Ion Accelerators for Advanced Radiation Application [TIARA], National Institute of Radiological Sciences [NIRS], Tsukuba, and Tohoku), Singapore, the United Kingdom, Australia, Croatia, Spain (CEMAM, Seville), Germany, France, the United States, and China. The layout of the scanning microbeam system at the NIRS, of the National Institutes for Quantum and Radiological Science and Technology (QST), is shown as an example in Fig. 4.13 [36].

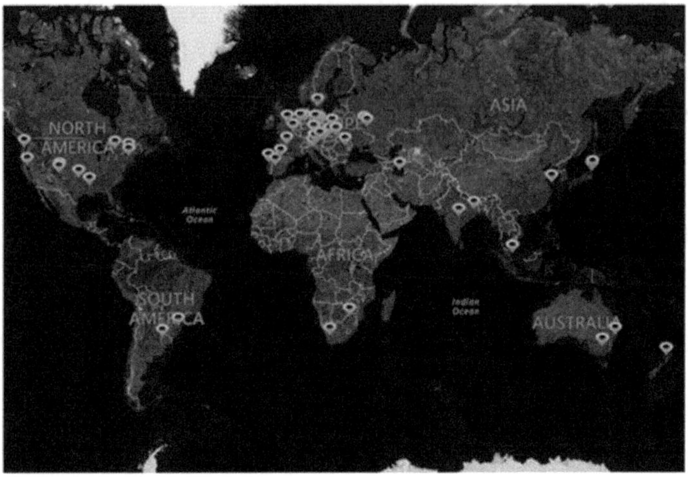

Figure 4.12 Ion microbeam facilities across the world. (https://nucleus.iaea.org/sites/accelerators).

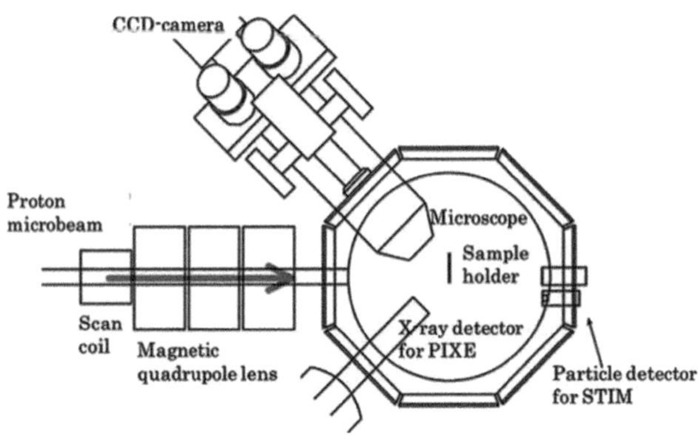

Figure 4.13 Layout of a scanning microbeam PIXE system at NIRS. Reprinted from Ref. [36], Copyright (2003), with permission from Elsevier.

Ions are usually classified by their mass numbers as light ions and heavy ions. Light ions usually refer to the ions with mass numbers of less than 4, for example, proton, deuterium ion (deuteron), and helium ion (α particle). Heavy ions refer to various ions of elements with mass numbers larger than 4. There is another classification: monoatomic ions, molecular ions, and cluster ions. The cluster ion beam is a group of large number of atoms ionized and accelerated as an ion beam. Interaction of the cluster ions with matter is very different from the interaction of the monoatomic ions. Many practical applications have been initiated due to their unique irradiation effects, especially on material surface modification by nuclear interaction with low-energy cluster ion beams. A cluster ion beam with MeV energy has been made available in recent years [38]. Clarification of the peculiar mechanisms of its interaction with matter and the applicability to the analysis technology are subjects of intensive research.

4.2.2 Accelerators for Ion Beam Analyses

In the electrostatic accelerators, charged particles are accelerated with an electrostatic high voltage of up to a few megavolts, generating continuous direct current (DC) ion beams. There are mainly two types of accelerators to obtain such high voltage. The first type, developed by Cockcroft and Walton [33], uses a voltage multiplying rectifier composed of stacks of diodes and capacitors (Fig. 4.14). The typical acceleration voltage is from hundreds of kilovolts to millions of volts. A rectifier circuit developed by Schenkel in 1919 [39] is also widely employed for high-voltage generation.

The second type, developed by Van de Graaff in 1931, places an electric charge on an insulator belt to carry the charge to an electrode of high voltage, up to ~20 MV (Fig. 4.15). Derived from this concept is a "pelletron," where a chain of metallic pellets connected by insulators is used to generate a high voltage, of over 25 MV. There is another type, called Disktron, invented by Isoya, where disks are used instead of the pellet chain [40].

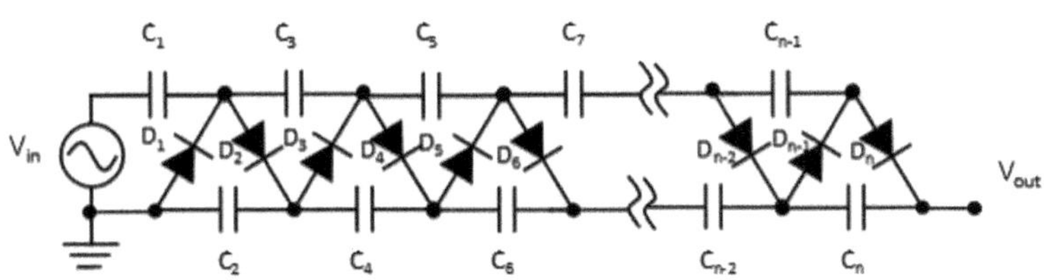

Figure 4.14 Schematic diagram of an n-stage Cockroft–Walton high-voltage multiplier.

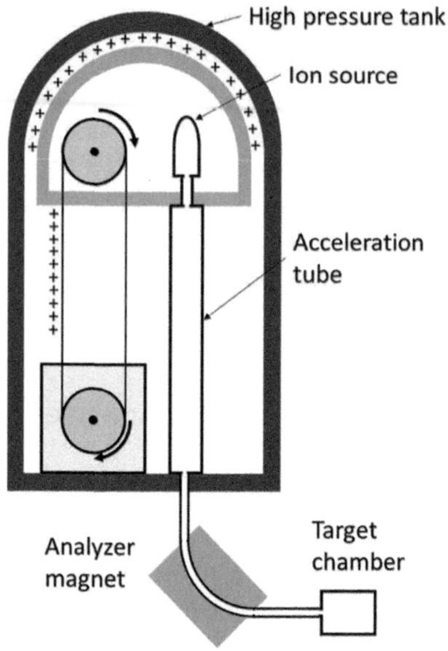

Figure 4.15 Schematic drawing of a single-ended accelerator with a Van de Graaff high-voltage generator. (A) A hollow metallic half-sphere mounted on the terminal. (B, C) Pulleys. A belt made of silk moves over the pulleys. C Is driven by a motor. (D, E) Two combs. D is maintained at a positive potential by a power supply. The upper comb E is connected to the inner side of the hollow metal sphere. The machine continuously transfers the positive charge to the sphere, and the potential of the sphere keeps increasing till it attains a maximum voltage. An ion source to generate a positive ion is mounted at the high-voltage terminal. Positive ions generated in the ion source are accelerated by the electric field along the acceleration tube. The accelerated beams are analyzed by the magnetic field of the analyzer magnet to obtain the required beam condition of the energy, mass, and charge.

There are two methods to accelerate the ions using an electrostatic field: single ended and tandem. In both methods, ions are extracted from the ion source and accelerated in the beam transport space called the "acceleration tube." The acceleration tube consists of the insulator tubes and the metal rings, with which the accelerating voltage is divided and impressed on each pole. The acceleration tube and the high-voltage terminal are installed inside a tank filled with high-pressure insulation gas (sulfur hexafluoride, SF_6) in order to maintain the high voltage on the terminal electrode of the accelerator. Positively charged ions are accelerated with the single-ended accelerator.

In the tandem accelerator, ions are accelerated to higher energies using two acceleration tubes connected to a high-voltage terminal, as shown in Fig. 4.16. In this scheme, negative ions generated by the negative ion source are accelerated by the positive high-voltage terminal through the first acceleration tube. Then the electrons are stripped with a film or gas at the high-voltage terminal so that negative ions are converted to positive ions. The positive ions thus generated are accelerated once again to the ground potential electrode through the second acceleration tube. Since the (negative) ion source is located outside of the high-pressure tank at the electrically ground level in this case, it is easier to change the ion species to be accelerated by changing the parameters of the ion source or the ion source itself.

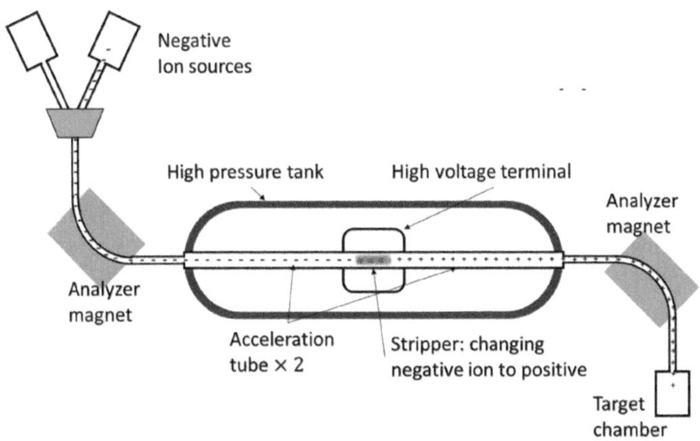

Figure 4.16 Schematic drawing of the tandem accelerator. Negative ions are accelerated in the first acceleration tube to the high-voltage terminal, changed to positive ions by the stripper gas or film, and accelerated again in the second tube to the earth level. Multiple ion sources can be connected.

There are various types of positive and negative ion sources for single-ended and tandem accelerators, respectively. The positive ion sources are generated by ionization in plasma discharges, which are classified into the DC type, the high-frequency type, and the microwave type. For lower-energy machines with high-voltage terminals of several hundred kilovolts, such as typical ion implanters, an ion source can be installed at the terminal in the atmosphere. All types of ion sources can be used in this case, since it is easy to change the ion species. In the case of MV-class single-ended machines, the high-voltage terminal is located in a closed tank filled with high-pressure SF_6 gas. The ion species that can be used with this type of machine are limited typically to hydrogen and helium, which are ionized by a high-frequency field at a relatively low power. In the case of a tandem accelerator, more than one negative ion sources can be connected to the injection beam line of the accelerator, as shown in Fig. 4.16. Although electron affinity must be positive to generate the negative ions, quite a wide range of ion species, including atoms, molecules, and clusters, can be accelerated.

While electrostatic accelerators are mostly used for ion beam analysis, small cyclotrons of ion beams of several MeVs of energy are used in some cases [38, 41, 42]. The basic structure of the classic cyclotron is shown in Fig. 4.17. A pair of electrodes with symmetrical semicircular shapes, called dee-electrodes, are placed in a uniform magnetic field. The ions generated at the central region of the cyclotron are accelerated repeatedly by high-frequency electric field at the gap of the dee-electrodes. This scheme works since the rotating period of the ion is constant (isochronous), because the orbital radius increases in proportion to the energy of the ions (in the nonrelativistic case). For higher-acceleration energies, where the relativistic effect cannot be ignored (over ~20 MeV for protons), AVF (azimuthally varying field) cyclotrons and ring cyclotrons have been developed for nuclear physics experiments such as exploration of new isotopes and exotic nuclei [43]. Also extra-large-scale particle accelerators have been constructed for particle physics [44, 45].

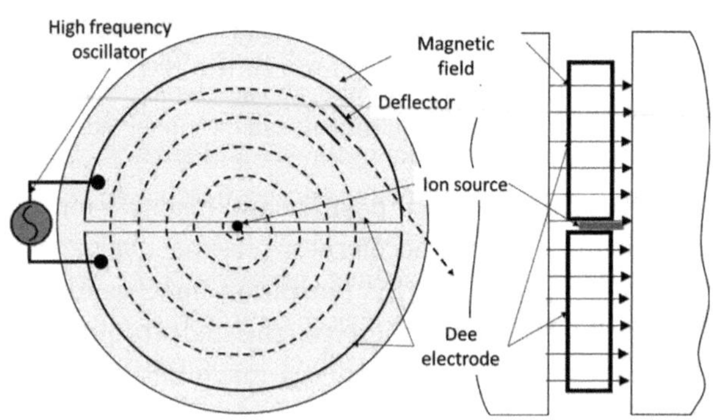

Figure 4.17 Schematic drawing of a classic cyclotron (Lawrence's patent in 1934).

4.2.3 Ion Microbeam Technology

An ion microbeam is an ion beam typically with energy of more than 1 MeV/nucleon and a beam size of less than 10 μm at the target. With the focused ion beam (FIB), spatial resolution of tens of nanometer has been achieved by ion beams of hundreds of kilo-electron-volt energy. FIBs are used for mask-less exposure with high spatial resolution of nanofabrication semiconductors [46, 47].

The microbeam was first developed in England in the 1970s as a proton microprobe for spatially resolved elemental analysis with ion beam analysis techniques such as PIXE analysis [48–50]. Since then, the ion microprobe has been utilized for biology, medicine, environment, geology, semiconductor industry, and other fields, with electrostatic accelerators of 2–4 MeV [51–54].

A typical ion microbeam system consists of a beam line, which is composed of a drift space, collimators for defining the initial beam size and divergence in the horizontal and vertical directions, a focusing system by combination of multiple quadrupole magnets, and an end station for bombardment of the target. Ion beam optics calculations are required for controlling the dynamics of the charged particles in the electromagnetic field to optimize the geometry and the magnetic field parameters of the system. The beam optics calculation is usually performed using a first- and second-order matrix multiplication computer program such as TRANSPORT [55, 56], which is intended for the design of static-magnetic beam transport systems.

Figure 4.18 shows an example of the beam optics calculation for the heavy-ion microbeam system at TIARA, QST [57]. The beam trajectories in the horizontal (x) and vertical (y) directions from the microslit to the target point are shown when a doublet quadrupole lens is placed near the target. The ratio between the drift lengths from the microslit to the lens (object) and the lens to the target (image) determines the magnification factor. A smaller beam can be obtained at the target with a larger object:image ratio. The stability in the energy and intensity of the ion beam is also required to obtain high spatial resolution.

As an example of the focusing parameters, the magnification factors and the achromatic aberration coefficients in the x and y planes are shown in the inset of Fig. 4.18. The achromatic aberration increases with the beam divergence (θ and φ) and the momentum spread (δ) of the beam. The inset table shows that the achromatic coefficient is approximately 100 μm/mrad%. The maximum beam divergence is 0.1 mrad at the largest beam diameter of 1 mm, where the quadrupole magnetic field acts correctly. In this case, δ should be 10^{-2}% at most in order to suppress the chromatic aberration to ~1/10 of the beam diameter. Usually RF-type accelerators like cyclotrons are not used for generating microbeams, due to wider energy variation and larger momentum spread of the beam, which is on the order of 10^{-1}%. However, there have been several facilities in which microbeam focusing systems have been developed on cyclotron beam lines [58, 59].

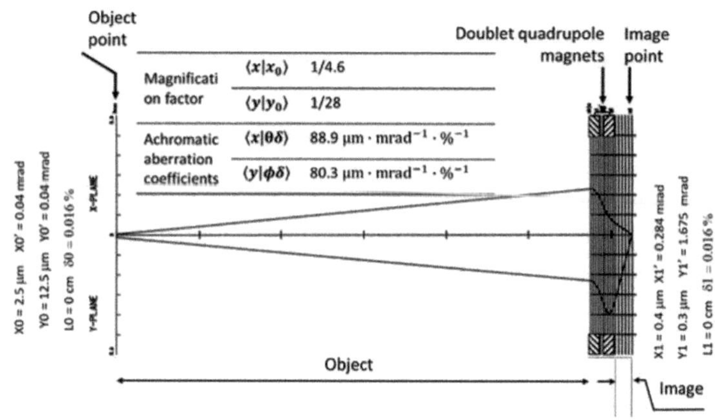

Figure 4.18 Beam optics for the microbeam system from the object point to the image point, which is the target point. Beam trajectories for horizontal (upper half) and vertical (lower half) directions are shown. The table in the inset shows focusing parameters, including achromatic aberration, for each direction.

In addition to the analytical uses of light-ion beams, heavy-ion beams have unique possibilities in terms of the use of the microbeam. The studies of radiation responses of semiconductor devices and biological cells require ion beams with a wide range of energy and mass. In the 1990s began the development of heavy-ion microbeam systems. A heavy-ion microbeam system was developed at GSI, Germany, to focus high-energy (over MeV/u) heavy ions and extract single ions at the end station [60, 61]. The Sandia National

Laboratory developed a heavy-ion microbeam system on a beam line of an electrostatic accelerator for testing microelectronic devices for space applications [62].

When semiconductor devices are exposed to high-energy particles in space environment, high-density charges of electrons and holes are generated by a single-ion injection into the device. This phenomenon, called single-event phenomena (SEP), has been studied by observing ion beam–induced charge (IBIC) signals or transient current shapes using the ion microbeam systems [63, 64]. This analytical technique is a powerful tool to reveal the behavior of the charges generated by energetic particle hits, such as dependencies on the structure and the hit position in the microelectronic devices. The SEP is also studied by measuring high-speed transient current pulses induced by single-ion bombardment of the device using a high-speed digital oscilloscope. This method is called the transient ion beam–induced current (TIBIC) or time-resolved IBIC (TRIBIC). These works have been successfully performed using heavy ions, of several MeVs/nucleon energy, that have a range of ~10 μm in semiconductor devices. Heavy ions of higher energies with longer ranges in the device are required to examine SEP in the deep layer of the device.

Such high-energy machines are also required for biological cell irradiation, because the ion microbeam has to be transmitted through a thin membrane window in order to irradiate the living samples under an atmospheric environment. The ions have to furthermore travel through several tens of micrometer thick layer of water in order to introduce radiation effect inside the biological cells and tissues. For testing high-energy deposition by various single heavy ions, MeV/u ion accelerators are required. It is difficult to generate heavy element microbeams with MeV/u energy using electrostatic accelerators, although there is a way to use a tandem accelerator with a high terminal voltage, like SNAKE at Munich [65]. In some cases, it is better to use the cyclotron, which can accelerate different ions with the same ratio of mass and charge almost under the same conditions [66]. In the single-ion irradiation to biological cells, it is necessary to determine when and where each ion has hit the individual cell. Secondary electrons, scintillation, and IBIC signals have been measured by single-ion irradiation [67].

Another major application of the MeV-energy ion microbeams is ion beam lithography to fabricate micro-/nanostructures with a high aspect ratio. Frank Watt at Singapore National University has established this proton beam writing technology using proton beams of a few MeVs with a spatial resolution that is better than 100 nm [68]. Other groups have also started to develop this technique in the last couple of decades [69, 70]. While a 2–3 MeV proton beam is suitable to expose photoresists, in a way similar to ultraviolet or electron beams, other ion beams with different species and energies can also be a source of exposure, since they can introduce sufficient energy deposition in the material. For example, a single particle with high linear energy transfer introduces enough energy deposition to decompose polymers along the ion track to create an etch pit. A so-called etch-pit membrane of a polymer such as CR-39 can be produced in this way [71]. This method is used for the evaluation of the position accuracy of a single ion hitting at the microbeam system [72]. Another application is radiation-induced cross linking of polymers, creating nanowires after development. The group of Seki at Osaka University has developed a single-particle nanofabrication technique and successfully demonstrated nanowire fabrication from polymers, including proteins, using the AVF cyclotron at TIARA [73, 74].

4.3 Ion Microbeam Analysis Technique

4.3.1 Micro-PIXE/Micro-PIGE Analysis

When a high-energy ion beam is focused down to several micrometers in diameter, it is possible to measure the spatial distribution of various elements by recording an X-ray spectrum in combination with the spatial information of the scanning microbeam on the sample. The PIXE analysis method using this type of ion microbeam is called a micro-PIXE analysis.

In the 2D scan of a microbeam, the detected signal is collected as the X-ray energy data E together with a set of beam scan position data (x, y) with a computer system equipped with the data input/output interface modules. The collected signal data are processed for spectrum analysis to construct the 2D elemental distributions.

The microbeam is scanned by a scanner consisting of a set of magnetic coils or electrostatic deflectors. The scanning region is within the reaches of the focusing lens system to reduce the risk of irregular focusing in the off-axis region of the quadrupole magnetic field at the large scanning angle. A sample can be also scanned using X/Y stages instead of beam scanning. An X/Y stage can be used for elemental mapping over a larger scale area than the normal beam scanning area or for direct access to the region of interest in the larger sample.

It is necessary to synchronize beam scanning or sample scanning with signal detection and data acquisition. When beam scanning and data collection are made in the same system, the positional data can be given internally at the E data collection timing. The detector identification is also required in addition to the data set of (x, y, E) when multiple detectors are used. Detection time information is also necessary if synchronization between detection signals is important.

The number of pixels in an imaging area, which is determined by the microbeam scanning step, defines the resolution of the image. Although a higher-resolution image can be obtained with a larger number of image pixels, longer measuring time is needed to acquire enough number of data per pixel for statistical accuracy. This also applies to the number of channels of the analog-to-digital converter (ADC), which corresponds to the energy resolution of the detector. In both cases, a larger memory capacity of the computer system and a higher processing speed are needed.

The detection efficiency depends on the solid angle of the detector. In covering the same total detection solid angle, increasing the number of detectors has merit than using a single detector of a large solid angle. Since a certain processing time is required for every signal detection in a single detector, the detection dead time significantly increases when the counting rate exceeds the limit determined by the dead time. This limit of the counting rate increases when multiple detectors are used.

However, multiple simultaneous detections in multiple detectors by one incident γ-ray can also occur due to Compton scattering. The possibility of multiple detection increases according to the number of detectors. It is necessary to build a coincidence process or algorithm in which the signals are excluded or summed when signals are detected by more than one detector within a short period. Owing to the diversity of interaction of the ion beam with the material, it is possible to analyze various kinds of interactions at the same time by use of an ion microbeam. For example, both the ionization and nuclear reaction analyses can be realized at the same time. A highly optimized arrangement for each detection system in a compact end station is required to utilize this feature effectively.

Figure 4.19 shows a configuration of the microbeam line comprising a three-slit system (microslit, MS; divergence-defining slit, DS; and baffle slit, BS) and a quadrupole doublet lens (DQ). When the object distance (from the MS to the entry of the DQ) and the image distance (from the exit of the DQ to the target) are 8.1 m and 0.2 m, respectively, a second-order TRANSPORT calculation shows that magnification factors in *X* and *Y* directions are respectively 1/12 and 1/60 in the case of 3 MeV H^+ [75]. The system usually has a two-axis electrostatic (or magnetic) scanner downstream (or upstream) of DQ, and the scan area is typically less than 1 mm × 1 mm.

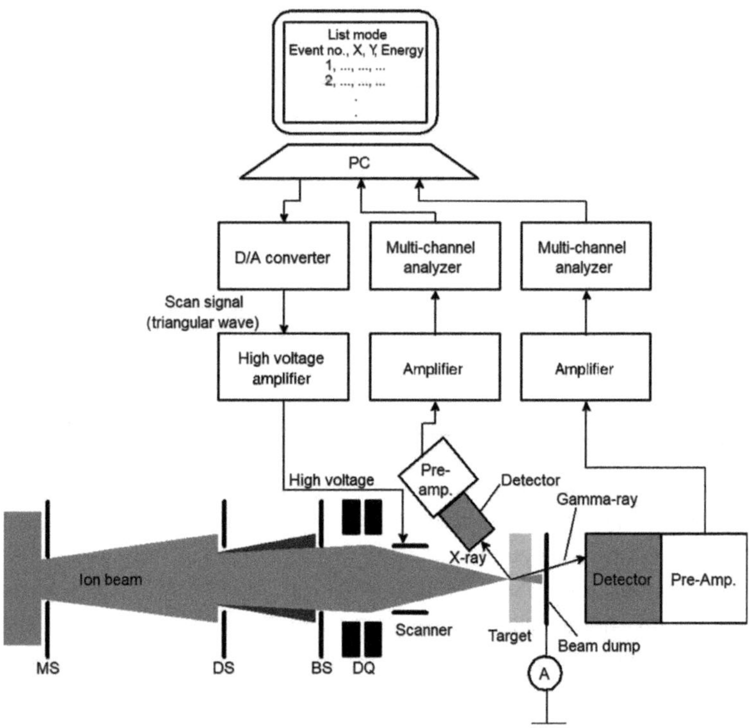

Figure 4.19 A schematic diagram of the micro-PIXE/micro-PIGE analyzer.

An in-air micro-PIXE analyzer was initially developed for elemental analysis in small biological samples [76]. The micro-PIXE system at TIARA consists of the light-ion microbeam line, a beam position control, and several X-ray detectors. The microbeam scans an interesting region of a target at the rate of several seconds per scan during measurement. The energy of the X-ray photons and the beam position are simultaneously recorded event by event and stored in list mode. The cross-sectional view of the in-air micro-PIXE system is shown in Fig. 4.20. The microbeam passes through a 5 μm thick polyethylene terephthalate membrane and irradiates the target in air. The transmitted beam is stopped in a beam dump made of glassy carbon.

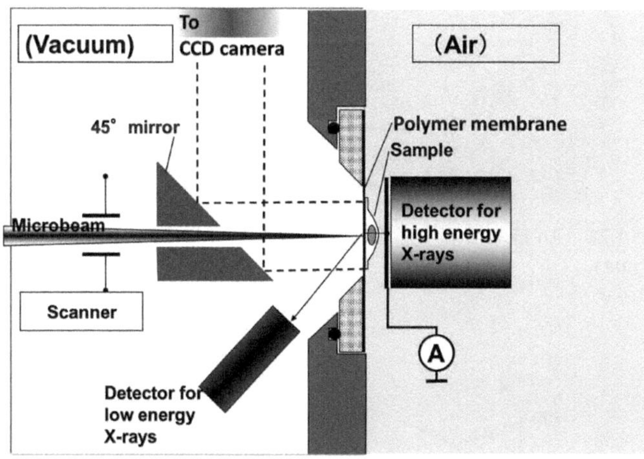

Figure 4.20 A cross-sectional view of the in-air micro-PIXE system.

4.3.2 Microbeam Analyses at the Ion Beam Irradiation Facility of QST

TIARA (Fig. 4.21), located at Takasaki Advanced Radiation Research Institute, QST, formerly Japan Atomic Energy Research Institute (JAERI), since 1993, is a unique ion accelerator facility dedicated mainly to ion beam applications on material science and biotechnology [77]. For utilizing unique characteristics of the ion beam in a wide field of science, TIARA was designed as an accelerator complex consisting of an AVF cyclotron, three electrostatic 3 MV tandem accelerators, a 3 MV single-ended accelerator, and a 400 kV ion implanter, together with a variety of beam transport systems and target chambers for performing ion beam irradiation experiments with a wide range of parameters [78, 79]. They cover ion species from hydrogen to bismuth and total energy from 20 keV to 620 MeV, as shown in Fig. 4.22 [80].

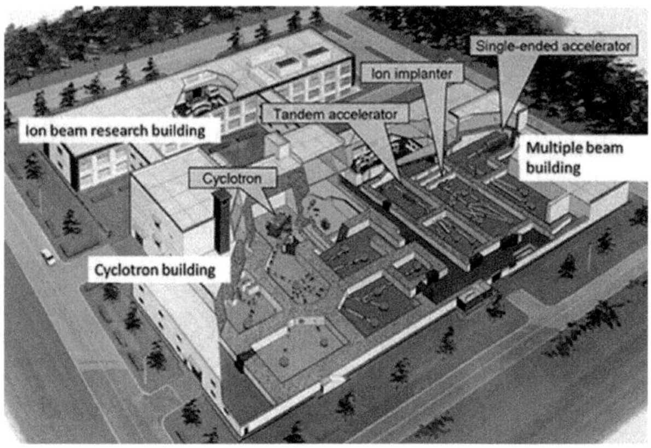

Figure 4.21 An illustration of TIARA buildings with the accelerators and the beam lines.

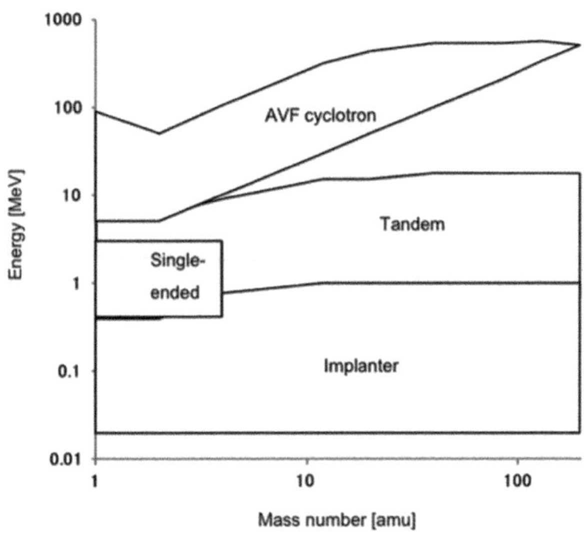

Figure 4.22 The range of the ion species and the beam energy covered by the four accelerators at TIARA.

The ion microbeam is one of the novel technologies developed for TIARA in 1990 [81]. It can control the position and area of the target for ion bombardment with a high-spatial resolution of a micrometer or less. A 2D distribution of specific elements can be measured, and arbitrary micro-/nanopatterns can be drawn to introduce micro/ nanostructures by scanning the microbeam on the sample. Three different types of focusing microbeam systems have been developed at TIARA: the light-ion microbeam system for the 3 MV single-ended electrostatic accelerator [82], the heavy-ion system for the 3 MV tandem machine [57], and the high-energy heavy-ion system for the AVF cyclotron ($K = 110$) [83].

While they all have high spatial resolution (more than 1 μm), the light-ion microbeam system in particular was designed to achieve the highest spatial resolution for microbeam analysis such as micro-PIXE [81]. In this case, an accelerator of extremely high voltage stability (better than $\pm 1 \times 10^{-5}$) is required in order to reduce chromatic aberration in beam focusing. Figure 4.23 shows a 3 MV single-ended accelerator (NC3000B, Nissin High Voltage Co.) with a balanced-Schenkel-type DC power supply having a voltage-stability of $\pm 1 \times 10^{-5}$ is installed for the TIARA light-ion microbeam system. The accelerator provides beams of H^+, D^+, and He^+ generated by an RF ion source with energy ranging from 0.4 MeV to 3 MeV. The RF ion source is installed on the high-voltage terminal with three gas bottles, hydrogen, helium, and deuteron.

Figure 4.23 Photograph of a 3 MV single-ended accelerator at TIARA.

The light-ion microbeam systems was established in 1995 as the second microbeam system after the first one, of the heavy-ion microbeam system. In the design of the focusing lens, a simple doublet quadrupole magnet lens configuration was adopted so as to reduce the influence of the parasitic aberration caused by the location and inclination of and the revolving misalignment between the lenses. The object and the image distances are 8.1 m and 0.2 m, respectively, with the magnification factors in *X* and *Y* directions of 1/12 and 1/60. A pair of X and Y electrostatic scanner lies downstream of a doublet, and the scan area is typically 800 μm × 800 μm in the case of 3 MeV H^+. The highest spatial resolution, determined by the beam size, of 0.20 μm, has been achieved. This beam size was estimated by measuring the full width at half maximum of the secondary electron yield curves for *X* and *Y* directions at the sharp edges, as shown in Fig. 4.24.

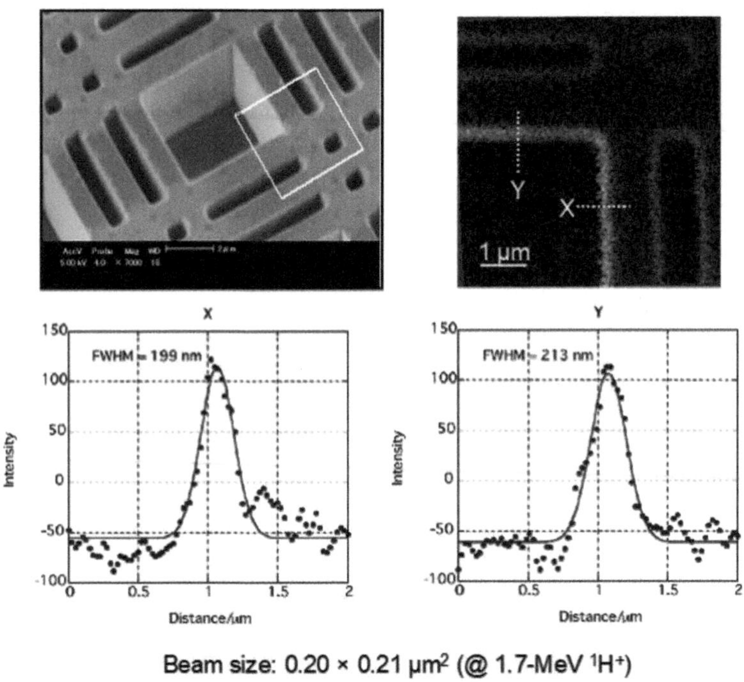

Figure 4.24 Beam size measurement using secondary electron imaging. The secondary electron yield curves for sharp edges in *X* and *Y* directions were fit with Gaussian curves and the full width at half maximum was obtained as the beam size.

This micro-PIXE analyzing system was initially developed for elemental analysis in small biological samples [82, 83]. Biomedical applications have been performed with the samples in the atmosphere, which are in contact with the beam exit window of a thin polymer film. The PIXE, PIGE, and ion luminescent analyses were performed simultaneously depending on the object elements or the chemical states. This system was also used effectively in the analyses of aerosols and lithium-ion battery materials. For those samples, the region of interest can be positioned at the central area of the target window, with a diameter of 1 mm, thus covering the microbeam scan area of 800 μm × 800 μm.

4.4 Summary

Owing to the diversity of ion beam interaction with the material, an ion beam can be utilized for a wide variety of analyses, such as PIXE, NRA, PIGE, RBS, and SIMS, in the same end station, even at the same time. In this chapter, each analytical method has been introduced. The basics for ion beam interaction with material, accelerator technologies, and microbeam as a novel ion beam technology are reviewed. Detailed descriptions are given on the microbeam analysis system, comprising microbeam scanning and data processing of the detection signals to visualize elemental distributions in the sample. TIARA is the first ion accelerator facility dedicated to ion irradiation applications. The ion microbeam is one of the representative technologies in TIARA for microbeam analysis in various applications.

References

1. Bozoin, M., Tesmer, J. and Nastasi, M. (1995). *Handbook of Modern Ion Beam Analysis* (Materials Research Society, Pittsburgh, PA).
2. Sera, K., Futatsugawa, S. and Murao, S. (2002). Quantitative analysis of untreated hair samples for monitoring human exposure to heavy metals, *Nucl. Instrum. Methods Phys. Res., Sect. B*, **189**, pp. 174–179.
3. Kada, W., Satoh, T., Yokoyama, A., Koka, M. and Kamiya, T. (2013). Determining aerosol particles by in-air micro-IL analysis combined with micro-PIXE, *Nucl. Instrum. Methods Phys. Res., Sect. B*, **306**, pp. 94–98.

4. Legge, G. J. and Cholewa, M. (1994). The principles of proton probe microanalysis in biology, *Scanning Microsc. Suppl.*, **8**, pp. 295–315.
5. Breese, M. B. H., Grime, G. W. and Watt, F. (1992). The nuclear microprobe, *Annu. Rev. Nucl. Part. Sci.*, **42**, pp. 1–38.
6. Satoh, T., Yokoyama, A., Kitamura, A., Ohkubo, T., Ishii, Y., Takahatake, Y., Watanabe, S., Koma, Y. and Kada, W. (2016). Particle induced X-ray emission-computed tomography analysis of an adsorbent for extraction chromatography, *Nucl. Instrum. Methods Phys. Res., Sect. B*, **371**, pp. 419–423.
7. Michelet, C., Barberet, P., Moretto, P. and Seznec, H. (2015). Development and applications of STIM- and PIXE-tomography: a review, *Nucl. Instrum. Methods Phys. Res., Sect. B*, **363**, pp. 55–60.
8. Bethe, H. A. and Ashkin, J. (1953). In *Experimental Nuclear Physics* (John Wiley & Sons, Chapman & Hall, New York, London), p. 253.
9. Barkas, W. H., Dyer, J. N. and Heckman, H. H. (1963). Resolution of the Σ−-*mass* anomaly, *Phys. Rev. Lett.*, **11**, pp. 26–28.
10. Andersen, H. H., Simonsen, H. and Sørensen, H. (1969). An experimental investigation of charge-dependent deviations from the Bethe stopping power formula, *Nucl. Phys. A*, **125**, pp. 171–175.
11. ICRU Report No. 49. (1993). Stopping power and ranges for protons and alpha particles, International Commission on Radiation Units and Measurements, Bethesda, MD, USA.
12. Ziegler, J. F. (1998). RBS/ERD simulation problems: stopping powers, nuclear reactions and detector resolution, *Nucl. Instrum. Methods Phys. Res., Sect. B*, **136**, pp. 141–146.
13. Mayer, M. (1999). SIMNRA, a simulation program for the analysis of NRA, RBS and ERDA, *AIP Conf. Proc.*, **475**, pp. 541–544.
14. Johansson, S. A. E., Campbell, J. L. and Malmqvist, K. G. (1995). *Particle-Induced X-Ray Emission Spectrometry (PIXE)* (John Wiley & Sons).
15. Ishii, K., Morita, S., Tawara, H., Chu, T. C., Kaji, H. and Shiokawa, T. (1975). Quantitative trace-element analysis by proton-induced X-rays, *Nucl. Instrum. Methods*, **126**, pp. 75–80.
16. Thompson, A., Lindau, I., Attwood, D., Liu, Y., Gullikson, E., Pianetta, P., Howells, M., Robinson, A., Kim, K.-J., Scofield, J., Kirz, J., Underwood, J., Kortright, J., Williams, G. and Winick, H. (2009). *X-Ray Data Booklet* (Lawrence Berkeley National Laboratory).
17. Brandt, W. and Lapicki, G. (1981). Energy-loss effect in inner-shell Coulomb ionization by heavy charged particles, *Phys. Rev. A*, **23**, pp. 1717–1729.

18. Liu, Z. and Cipolla, S. J. (1996). ISICS: a program for calculating K-, L- and M-shell cross sections from ECPSSR theory using a personal computer, *Comput. Phys. Commun.*, **97**, pp. 315–330.

19. Batič, M., Pia, M. G. and Cipolla, S. J. (2012). ISICSoo: a class for the calculation of ionization cross sections from ECPSSR and PWBA theory, *Comput. Phys. Commun.*, **183**, pp. 398–404.

20. Schoonjans, T., Brunetti, A., Golosio, B., Sanchez del Rio, M., Solé, V. A., Ferrero, C. and Vincze, L. (2011). The xraylib library for X-ray–matter interactions. Recent developments, *Spectrochim. Acta, Part B*, **66**, pp. 776–784.

21. Ishii, K. and Morita, S. (1988). Theoretical estimation of PIXE detection limits, *Nucl. Instrum. Methods Phys. Res., Sect. B*, **34**, pp. 209–216.

22. West, H. I., Mustafa, M. G., Lanier, R. G. and O'Brien, H. (1991). Interpretation of ion-range recoil data obtained from activated-foil measurements of nuclear excitation functions, *Phys. Rev. C*, **43**, pp. 1352–1362.

23. Lanford, W. A., Trautvetter, H. P., Ziegler, J. F. and Keller, J. (1976). New precision technique for measuring the concentration versus depth of hydrogen in solids, *Appl. Phys. Lett.*, **28**, pp. 566–568.

24. Lanford, W. A. (1978). 15N hydrogen profiling: scientific applications, *Nucl. Instrum. Methods*, **149**, pp. 1–8.

25. Xiong, F., Rauch, F., Chengru Shi, Zhuying Zhou, Livi, R. P. and Tombrello, T. A. (1987). Hydrogen depth profiling in solids: a comparison of several resonant nuclear reaction techniques, *Nucl. Instrum. Methods Phys. Res., Sect. B*, **27**, pp. 432–441.

26. Clark, G. J., White, C. W., Allred, D. D., Appleton, B. R., Koch, F. B. and Magee, C. W. (1978). The application of nuclear reactions for quantitative hydrogen analysis in a variety of different materials problems, *Nucl. Instrum. Methods*, **149**, pp. 9–18.

27. Iwata, Y., Fujimoto, F., Vilata, E., Ootuka, A., Komaki, K., Kobayashi, K., Yamashita, H. and Murata, Y. (1987). Detection of adsorbed hydrogen on W(001) by using 1H(15N, αγ)12C reaction, *Jpn. J. Appl. Phys.*, **26**, p. L1026.

28. Hjörvarsson, B. and Rydén, J. (1990). Off-resonance and straggling measurements using the 1H(15N, αγ)12C reaction, *Nucl. Instrum. Methods Phys. Res., Sect. B*, **45**, pp. 36–40.

29. Fallavier, M., Hjörvarsson, B., Benmansour, M. and Thomas, J. P. (1992). Determination of hydrogen surface coverage of $Pt_{0.5}Ni_{0.5}$ single crystals by NRA, *Nucl. Instrum. Methods Phys. Res., Sect. B*, **64**, pp. 83–87.

30. Lanford, W. A. (1992). Analysis for hydrogen by nuclear reaction and energy recoil detection, *Nucl. Instrum. Methods Phys. Res., Sect. B*, **66**, pp. 65–82.

31. Ziegler, J. F., Wu, C. P., Williams, P., White, C. W., Terreault, B., Scherzer, B. M. U., Schulte, R. L., Schneid, E. J., Magee, C. W., Ligeon, E., 'Ecuyer, J. L., Lanford, W. A., Kuehne, F. J., Kamykowski, E. A., Hofer, W. O., Guivarc'h, A., Filleux, C. H., Deline, V. R., Evans, C. A., et al. (1978). Profiling hydrogen in materials using ion beams, *Nucl. Instrum. Methods*, **149**, pp. 19–39.

32. Van de Graaff, R. J. (1931). A 1,500,000 volt electrostatic generator, *Phys. Rev.*, **38**, pp. 1919–1920.

33. Cockcroft, J. D. and Walton, E. T. S. (1932). Experiments with high velocity positive ions. (I) Further developments in the method of obtaining high velocity positive ions, *Proc. R. Soc. London, Ser. A*, **136**, pp. 619–630.

34. Ising, G. (1924). Prinzip einer methode zur herstellung von kanalstrahlen hoher voltzahl, *Ark. Mat. Astr. Fys.*, **18**, pp. 1–4.

35. Lawrence, E. O. and Livingston, M. S. (1932). The production of high speed light ions without the use of high voltages, *Phys. Rev.*, **40**, pp. 19–35.

36. Imaseki, H., Yukawa, M., Watt, F., Ishikawa, T., Iso, H., Hamano, T., Matsumoto, K. and Yasuda, N. (2003). The scanning microbeam PIXE analysis facility at NIRS, *Nucl. Instrum. Methods Phys. Res., Sect. B*, **210**, pp. 42–47.

37. Isoya, A., Miyake, Y., Takagi, K., Uchiyama, T., Yui, K., Kikuchi, R., Komiya, S. and Hayashi, C. (1985). 1 MV rotating-disc type high voltage generator for application to an implanter, *Nucl. Instrum. Methods Phys. Res., Sect. B*, **6**, pp. 250–257.

38. Choi, Y.-G. and Kim, Y.-S. (2015). External beam's nozzle design for the CRC cyclotron PIXE/PIGE, *J. Korean Phys. Soc.*, **66**, pp. 394–398.

39. Schenkel, M. (1919). New method of producing high d.c. pressures, *Electrotechnische Zeitschrift ETZ*, **40**, pp. 333–334.

40. Isoya, A., Kobayashi, K., Nakashima, T. and Maeda, T. (1985). Rotating-disc type high voltage generator, *Nucl. Instrum. Methods Phys. Res., Sect. A*, **236**, pp. 215–221.

41. Quaedackers, J. A., de Voigt, M. J. A., Mutsaers, P. H. A., de Goeij, J. J. M. and van der Vusse, G. J. (1999). Nuclear elemental analyses with a cyclotron on biomedical samples, *Nucl. Instrum. Methods Phys. Res., Sect. B*, **156**, pp. 252–257.

42. Futatsugawa, S., Hatakeyama, S., Saitou, Y. and Sera, K. (1996). Improvement on experimental procedure of various samples in PIXE analysis, *Nucl. Instrum. Methods Phys. Res., Sect. B*, **109**, pp. 113–120.

43. Lee, S. Y. (2012). *Accelerator Physics*, 3rd ed. (World Scientific Publishing, Hackensack, NJ).

44. Tanaka (Konstantinova), O. (2017). Review of the elementary particles physics in the external electromagnetic fields studies at KEK, *Russ. Phys. J.*, **59**, pp. 1864–1869.

45. Pancheri, G. and Srivastava, Y. N. (2017). Introduction to the physics of the total cross section at LHC - a review of data and models, *Eur. Phys. J. C*, **77**, p. 150.

46. Moore, D. F., Daniel, J. H. and Walker, J. F. (1997). Nano-and micro-technology applications of focused ion beam processing, *Microelectron. J.*, **28**, pp. 465–473.

47. Volkert, C. A. and Minor, A. M. (2007). Focused ion beam microscopy and micromachining, *MRS Bull.*, **32**, pp. 389–399.

48. Cookson, J. A., Ferguson, A. T. G. and Pilling, F. D. (1972). Proton microbeams, their production and use, *J. Radioanal. Nucl. Chem.*, **12**, pp. 39–52.

49. Legge, G. J. F., McKenzie, C. D., Mazzolini, A. P., Sealock, R. M., Jamieson, D. N., O'Brien, P. M., McCallum, J. C., Allan, G. L., Brown, R. A., Colman, R. A., Kirby, B. J., Lucas, M. A., Zhu, J. and Cerini, J. (1986). Microbeam imaging at micron and submicron resolution, *Nucl. Instrum. Methods Phys. Res., Sect. B*, **15**, pp. 669–674.

50. Grime, G. W., Dawson, M., Marsh, M., McArthur, I. C. and Watt, F. (1991). The Oxford submicron nuclear microscopy facility, *Nucl. Instrum. Methods Phys. Res., Sect. B*, **54**, pp. 52–63.

51. Watt, F. (1995). Nuclear microscopy in the life sciences, *Nucl. Instrum. Methods Phys. Res., Sect. B*, **104**, pp. 276–284.

52. Pallon, J., Knox, J., Forslind, B., Werner-Linde, Y. and Pinheiro, T. (1993). Applications in medicine using the new Lund microprobe, *Nucl. Instrum. Methods Phys. Res., Sect. B*, **77**, pp. 287–293.

53. Grime, G. W. and Watt, F. (1993). A survey of recent PIXE applications in archaeometry and environmental sciences using the Oxford scanning proton microprobe facility, *Nucl. Instrum. Methods Phys. Res., Sect. B*, **75**, pp. 495–503.

54. Ryan, C. G. and Griffin, W. L. (1993). The nuclear microprobe as a tool in geology and mineral exploration, *Nucl. Instrum. Methods Phys. Res., Sect. B*, **77**, pp. 381–398.

55. Brown, K. L. (1982). A first- and second-order matrix theory for the design of beam transport systems and charged particle spectrometers, *SLAC Rep.*, **75**, pp. 71–134.

56. Brown, K. L., Carey, D. C., Iselin, C. and Rothacker, F. (1980). Transport: a computer program for designing charged particle beam transport systems, *CERN Yellow Reports, CERN*-80-04.

57. Kamiya, T., Utsunomiya, N., Minehara, E., Tanaka, R. and Ohdomari, I. (1992). Microbeam system for study of single event upset of semiconductor-devices, *Nucl. Instrum. Methods B*, **64**, pp. 362–366.

58. Quaedackers, J. A., de Voigt, M. J. A., Mutsaers, P. H. A., de Goeij, J. J. M. and van der Vusse, G. J. (1999). Nuclear elemental analyses with a cyclotron on biomedical samples, *Nucl. Instrum. Methods Phys. Res., Sect. B*, **156**, pp. 252–257.

59. Koyama-Ito, H., Wada, E. and Ito, A. (1987). A cyclotron microbeam and its application to biomedical samples, *Nucl. Instrum. Methods Phys. Res., Sect. B*, **22**, pp. 205–209.

60. Fischer, B. E. (1991). Single-particle techniques, *Nucl. Instrum. Methods Phys. Res., Sect. B*, **54**, pp. 401–406.

61. Fischer, B. E. and Metzger, S. (1995). Aiming and hit verification in single ion techniques, *Nucl. Instrum. Methods Phys. Res., Sect. B*, **104**, pp. 7–12.

62. Sexton, F. W., Horn, K. M., Doyle, B. L., Laird, J. S., Cholewa, M., Saint, A. and Legge, G. J. F. (1993). Ion-beam-induced charge-collection imaging of CMOS ICs, *Nucl. Instrum. Methods Phys. Res., Sect. B*, **79**, pp. 436–442.

63. Hirao, T., Mori, H., Laird, J. S., Onoda, S., Wakasa, T., Abe, H. and Itoh, H. (2003). Studies on single-event phenomena using the heavy-ion microbeam at JAERI, *Nucl. Instrum. Methods Phys. Res., Sect. B*, **210**, pp. 227–231.

64. Medunić, Z., Jakšić, M., Pastuović, Ž. and Skukan, N. (2003). Temperature dependent TRIBIC in CZT detectors, *Nucl. Instrum. Methods Phys. Res., Sect. B*, **210**, pp. 237–242.

65. Hauptner, A., Dietzel, S., Drexler, G. A., Reichart, P., Krücken, R., Cremer, T., Friedl, A. A. and Dollinger, G. (2004). Microirradiation of cells with energetic heavy ions, *Radiat. Environ. Biophys.*, **42**, pp. 237–245.

66. Kurashima, S., Okumura, S., Miyawaki, N., Kashiwagi, H., Satoh, T., Kamiya, T., Fukuda, M. and Yokota, W. (2013). Short-time change of heavy-ion microbeams with different mass to charge ratios by scaling method for the JAEA AVF cyclotron, *Nucl. Instrum. Methods Phys. Res., Sect. B*, **306**, pp. 40–43.

67. Kamiya, T., Satoh, T., Koka, M. and Kada, W. (2015). Development of microbeam technology to expand applications at TIARA, *Nucl. Instrum. Methods Phys. Res., Sect. B*, **348**, pp. 4–7.

68. Watt, F., Breese, M. B. H. and Bettiol, A. A. (2007). Proton beam writing, *Mater. Today*, **10**, pp. 20–29.

69. Rajta, I., Gómez-Morilla, I., Abraham, M. H. and Kiss, Á. Z. (2003). Proton beam micromachining on PMMA, Foturan and CR-39 materials, *Nucl. Instrum. Methods Phys. Res., Sect. B*, **210**, pp. 260–265.

70. Uchiya, N., Harada, T., Murai, M., Nishikawa, H., Haga, J., Sato, T., Ishii, Y. and Kamiya, T. (2007). Micro-machining of resists on silicon by proton beam writing, *Nucl. Instrum. Methods Phys. Res., Sect. B*, **260**, pp. 405–408.

71. Omichi, H., Yoshida, M., Asano, M., Nagaoka, N., Kubota, H., Katakai, R., Spohr, R., Reber, N., Wolf, A., Alder, G. M., Ang, V., Bashford, C. L. and Pasternak, C. A. (1997). Application of ion beams to synthesis of environmentally responsive track membranes, *Nucl. Instrum. Methods Phys. Res., Sect. B*, **131**, pp. 350–356.

72. Hamano, T., Hirao, T., Nashiyama, I., Sakai, T., Kamiya, T., Ishii, K. and Takebe, M. (1998). Diagnosis of the profile of the heavy-ion microbeam and estimation of the aiming accuracy of the single-ion-hit with using CR-39, *Radiat. Phys. Chem.*, **53**, pp. 461–467.

73. Seki, S., Saeki, A., Choi, W., Maeyoshi, Y., Omichi, M., Asano, A., Enomoto, K., Vijayakumar, C., Sugimoto, M., Tsukuda, S. and Tanaka, S. (2012). Semiconducting cross-linked polymer nanowires prepared by high-energy single-particle track reactions, *J. Phys. Chem. B*, **116**, pp. 12857–12863.

74. Asano, A., Omichi, M., Tsukuda, S., Takano, K., Sugimoto, M., Saeki, A. and Seki, S. (2012). Fabrication and arrangement of "clickable" nanowires by the single-particle nanofabrication technique, *J. Phys. Chem. C*, **116**, pp. 17274–17279.

75. Kamiya, T., Suda, T. and Tanaka, R. (1996). Sub-micron microbeam apparatus for high resolution materials analyses, *Nucl. Instrum. Methods Phys. Res., Sect. B*, **118**, pp. 447–450.

76. Ishii, K., Sugimoto, A., Tanaka, A., Satoh, T., Matsuyama, S., Yamazaki, H., Akama, C., Amartivan, T., Endoh, H., Oishi, Y., Yuki, H., Sugihara, S., Satoh, M., Kamiya, T., Sakai, T., Arakawa, K., Saidoh, M. and Oikawa, S. (2001). Elemental analysis of cellular samples by in-air micro-PIXE, *Nucl. Instrum. Methods Phys. Res., Sect. B*, **181**, pp. 448–453.

77. Tanaka, R., Saido, M., Nashiyama, I., Naramoto, H., Suwa, T., Morita, Y., Toraishi, A., Watanabe, H., Tani, M., Tajima, S. and Okada, S. (1998). TIARA annual report, *JAERI-Review*, 98-016.

78. Arakawa, K., Nakamura, Y., Yokota, W., Fukuda, M., Nara, T., Agematsu, T., Okumura, S., Ishibori, I., Karasawa, T. and Tanaka, R. (1992). Construction and first year's operation of the JAERI AVF cyclotron, *Proc. 13th Int. Conf. on Cyclotron and their Applications*, pp. 119–122.

79. Saitoh, Y., Tajima, S., Takada, I., Mizuhashi, K., Uno, S., Ohkoshi, K., Ishii, Y., Kamiya, T., Yotumoto, K., Tanaka, R. and Iwamoto, E. (1994). TIARA electrostatic accelerators for multiple ion beam application, *Nucl. Instrum. Methods Phys. Res., Sect. B*, **89**, pp. 23–26.

80. Kurashima, S., Satoh, T., Saitoh, Y. and Yokota, W. (2017). Irradiation facilities of the takasaki advanced radiation research institute, *Quantum Beam Sci.*, **1**, p. 2.

81. Kamiya, T. (2015). Microbeam systems at TIARA, *Int. J. PIXE*, **25**, pp. 135–146.

82. Kamiya, T., Suda, T. and Tanaka, R. (1995). Submicron microbeam apparatus using a single-ended accelerator with very high voltage stability, *Nucl. Instrum. Meth. B*, **104**, pp. 43–48.

83. Oikawa, M., Kamiya, T., Fukuda, M., Okumura, S., Inoue, H., Masuno, S., Umemiya, S., Oshiyama, Y. and Taira, Y. (2003). Design of a focusing high-energy heavy ion microbeam system at the JAERI AVF cyclotron, *Nucl. Instrum. Methods Phys. Res., Sect. B*, **210**, pp. 54–58.

Chapter 5

Ion Beam Analysis of Lithium-Ion Batteries

Akiyoshi Yamazaki,[a] Takahiro Satoh,[b] Kazuhisa Fujita,[c] Kunioki Mima,[c] and Yoshiaki Kato[c]
[a]*Faculty of Pure and Applied Sciences, University of Tsukuba, 1-1-1 Tennodai, Tsukuba, Ibaraki Prefecture 305-8577, Japan*
[b]*Takasaki Advanced Radiation Research Institute, Quantum Beam Science Research Directorate, National Institutes for Quantum and Radiological Science and Technology, 1233 Watanuki-machi, Takasaki, Gunma Prefecture 370-1292, Japan*
[c]*The Graduate School for the Creation of New Photonics Industries, 1955-1 Kurematsu-cho, Nishi Ward, Hamamatsu, Shizuoka Prefecture 431-1202, Japan*
yamazaki@tac.tsukuba.ac.jp

5.1 Application of Micro-PIGE and Micro-PIXE in Lithium-Ion Battery Diagnostics

Ion beam analysis is a material analysis method using ion accelerators as described in Chapter 4. This application was developed in the 1950s with the ion accelerators dedicated for nuclear physics by Van de Graaff. Ion beam analysis helps measure elemental concentration by using various kinds of interactions of the ion beam in a sample.

Lithium-Ion Batteries: Overview, Simulation, and Diagnostics
Edited by Yoshiaki Kato, Zenpachi Ogumi, and José Manuel Perlado Martin

ISBN 978-981-4800-40-2 (Hardback), 978-0-429-25934-0 (eBook)
www.panstanford.com

The following methods have been employed for ion beam analysis (see also Sections 4.1 and 4.2, Chapter 4).

- Rutherford backscattering spectroscopy (RBS)
- Elastic recoil detection analysis
- Nonresonant and resonant nuclear reaction analysis (NRA, RNRA)
- Particle-induced γ-ray emission (PIGE)
- Particle-induced X-ray emission (PIXE)

Reference [1] is a representative guide for ion beam analysis.

5.1.1 Introduction to PIGE and PIXE for Lithium-Ion Battery Diagnostics

In this section, we describe the development of PIXE and PIGE techniques for characterizing lithium-ion batteries (LIBs), which has been realized by two important breakthroughs: (i) a microbeam technology to analyze elemental distribution in a LIB and (ii) an external beam technology to irradiate the sample in an atmospheric or any gas-filled environment by a scanning microbeam. The PIXE and PIGE with microbeams, which we call micro-PIXE and micro-PIGE hereafter, are very useful for measuring elemental distribution in LIB electrodes on a microscale. These breakthroughs are effective for in situ measurements of LIBs in an inert gas environment to prevent lithium from reacting chemically in the atmosphere. In this chapter, we will focus on ion microbeam analysis of LIBs, with a submicrometer-diameter ion beam of several mega–electron volts energy and 100 pA–1 nA current [2].

Since PIGE is sensitive to light elements like lithium, the combination of PIGE and PIXE is suitable for imaging the distributions of lithium and other elements in the LIB . Application of this ion beam analysis technology to the LIB diagnostics is relatively new. The first application was made by Tadić et al. in Croatia in the 2000s [3, 4]. Then the ion microbeam technique was applied in Japan in 2012 by the Japanese and Spanish joint group to characterize LIBs. The composite electrode materials, $LiNi_{0.8}Co_{0.15}Al_{0.05}O_2$ (hereafter referred to as LNO) and $LiFePO_4$ (hereafter referred to as LFP), were investigated by the micro-PIGE and micro-PIXE technologies. As described in detail in the following sections, the electrodes in various charging conditions were diagnosed [5, 6].

In the proton beam diagnosis of lithium, the following nuclear reactions can be used: $^{7}Li(p, p'\gamma)^{7}Li$ emitting a γ-ray of 478 keV and $^{7}Li(p, n\gamma)^{7}Be$ emitting a γ-ray of 429 keV for PIGE. The latter reaction has a threshold of 1.8 MeV. For PIXE, we can use the X-ray lines of Ni-K_{α}/K_{β}, Fe-K_{α}/K_{β}, Co-K_{α}/K_{β}, P-K_{α}/K_{β}, and so on, which are in the range of 2–8 keV.

In this section, the first application of PIGE and PIXE to diagnostics of LIB electrodes is summarized and then some examples of LIB electrode diagnostics with micro-ion beam analyses (hereafter referred to as micro-IBA) at Takasaki Ion Accelerators for Advanced Radiation Application (TIARA), at the Takasaki Institute of the National Institutes for Quantum and Radiological Science and Technology (QST), are described.

5.1.2 Micro-PIGE and Micro-PIXE Diagnostics by an External Proton Beam

Tadić et al., of the Ruđer Bošković Institute in Croatia, proposed applying PIGE and PIXE to the characterization of a LIB [3, 4]. They measured elemental distributions of lithium and other elements contained in a polymer gel electrolyte by an external proton microbeam. A layout for the external beam PIXE/PIGE is shown in Fig. 5.1 [3]. In the Tadić experiments, a 4 MeV proton beam is transmitted through a beam port a few millimeters in diameter made of a 50 μm thick Kapton film and irradiates samples in the air. The PIGE γ-ray is measured by a γ-ray detector placed close to the sample, and the PIXE X-ray is measured by a Si(Li) detector in the vacuum chamber. Since the ion beam is scattered in the Kapton film and the air, the beam diameter on the sample surface was estimated to be about 80 μm for 4 MeV and 120 μm for 3 MeV protons, respectively, as shown in Fig. 5.2.

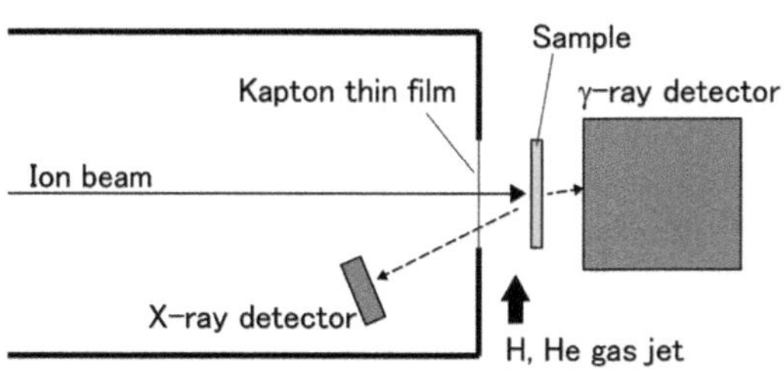

Figure 5.1 The layout for PIXE and PIGE measurements of LIB samples by an external beam [3].

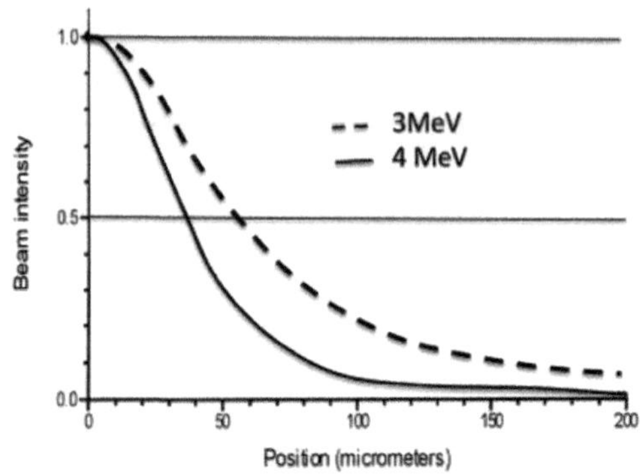

Figure 5.2 Proton beam spread due to multiple scattering in the Kapton film and the air. The half width at half maximum (HWHM) is 40 μm for 4 MeV and 60 μm for 3 MeV [3].

An external ion beam analysis is appropriate for measuring samples in normal pressure, such as uncovered wet samples. In the case of LIB samples, they contain a liquid electrolyte and cannot be measured inside the vacuum chamber, since the organic liquid electrolyte may boil in vacuum. Tadić et al. measured the elemental distributions in the cross sections of a polymer gel electrolyte layer in a LIB cell, which contains a mixture of lithium salts, organic solvent, and polymer gel [4].

The polymer gel electrolyte was a mixture of a polymer chain 94% of which was polyvinylidene fluoride (PVdF, $[CF_2CH_2]_n$) and 6% of which was a mixture of hexafluoropropylene (HFP, $[CF_2CFCF_3]_n$) and a lithium salt: $LiN(CF_3SO_2)_2$, called Li-imide, or $LiN(C_2F_5SO_2)_2$, called Libeti. Figure 5.3 shows the Li and F distribution in a cross section of the polymer gel electrolyte with Libeti, where the profiles along the cross section of 197 keV PIGE of F and 478 keV PIGE of Li are compared. These experiments demonstrate that PIGE is applicable to the nondestructive diagnostics of the nonuniform distribution of the Libeti concentration in the polymer chain of PVdF and HFP. This figure shows that the yield of 478 keV gamma of Li is reduced at the places of the higher yield of 197 keV gamma of F. This nonuniform distribution may degrade the LIB performance.

Tadić et al. measured Li, F, and other elements near the polymer gel (containing S and F) interfaces with the lithium anode and the spinel cathode ($LiMnO_2$ [LMO]) by PIGE for Li and F and by PIXE for S and Mn simultaneously at the IAB facility in Croatia offering a 3–4 MeV external proton beam [4].

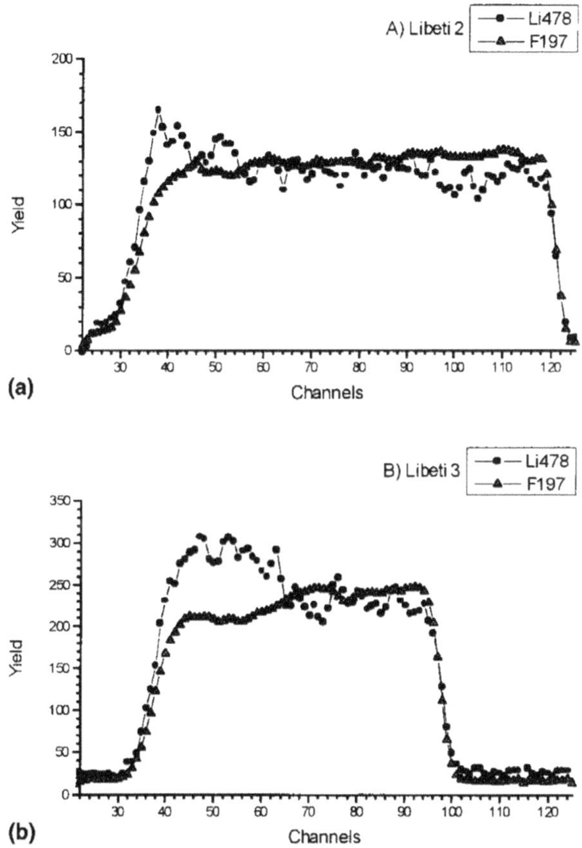

Figure 5.3 Distributions of the γ-ray yield for 478 keV of Li and 197 keV of F. (a) 1.51 mm thick Libeti [$LiN(C_2F_5SO_2)_2$] sample and (b) 0.8 mm thick Libeti sample. F is contained both in the polymer gel and in the lithium salt. Reprinted from Ref. [3], Copyright (2000), with permission from Elsevier.

In Fig. 5.4, the upper side is a lithium anode and the lower side is a polymer gel. The Li γ-ray yield increases locally, but with no correlation with the S X-ray signal. This may be due to the inhomogeneity of the anode. On the other hand, the S X-ray yield, which reflects the polymer gel concentration, is the highest at the interface, decreases sharply from the interface, and is uniform inside the lithium anode.

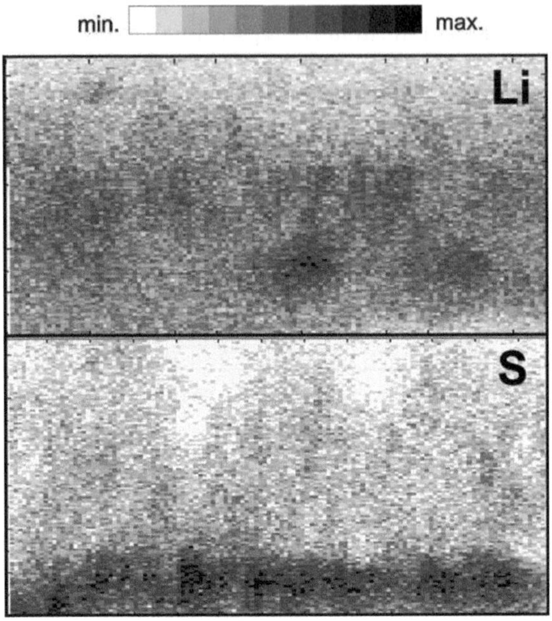

Figure 5.4 Li and S distributions near the Li-anode/polymer gel electrolyte interface. The scan size is 1 $mm^w \times 0.5$ mm^h. Reprinted from Ref. [4], Copyright (2001), with permission from Elsevier.

In Fig. 5.5, the S concentration is the highest near the interface and decreases deeper inside the LMO cathode (upper side of the figure). In the cathode, Li and Mn signals are correlated. In the higher area of Li/Mn signal, the S and F signals are lower. This means that the cathode has voids in LMO.

As shown above, the submillimeter-scale distributions of material inside LIB electrodes have been successfully measured.

5.1.3 Experimental Methods for Micro-PIGE and Micro-PIXE Diagnostics of Li-Ion Batteries in Vacuum

5.1.3.1 Introduction

Micrometer-scale distributions of Li and other elements were measured with the proton microbeam facility at TIARA, QST. In these experiments, LIB samples were set outside the vacuum chamber, as shown in Fig. 5.6. The LIB sample is used also for vacuum seal of the beam exit hole of the vacuum chamber. The experiments were carried out as Japan-Spain and Japan-Germany collaboration programs [5, 6]. The proton microbeam was used for measuring LIB electrodes of liquid or solid electrolyte. The characteristics of the proton microbeam are as follows:

- Beam diameter: ~1 μm
- Beam current: 100–500 pA
- Proton beam energy: 2–3 MeV
- X-ray detector: Si(Li)
- γ-ray detector: High-purity Ge (HPGe)

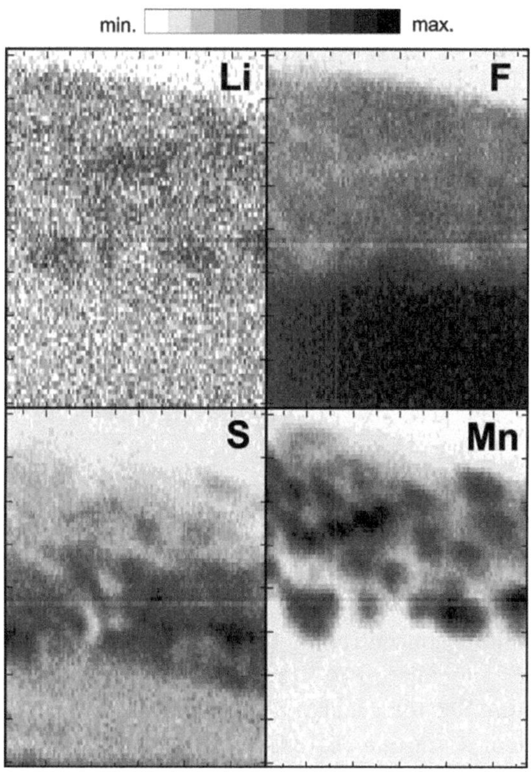

Figure 5.5 Li, F, S, and Mn distributions near the $LiMnO_2$ spinel cathode/polymer gel electrolyte interface. The scan size is 0.6 $mm^w \times 0.8\ mm^h$. Reprinted from Ref. [4], Copyright (2001), with permission from Elsevier.

The schematic view of the diagnostic system is shown in Fig. 5.6. Two kinds of positive electrode samples were fabricated.

- Liquid electrolyte LIB: (a) A mixture of active materials, LNO, LMO, $LiCoO_2$ (LCO), and $LiTiO_2$ (LTO), and (b) LFP, both with carbon black as the conductive carbon and a PVdF binder. Measurement was made with or without the liquid electrolyte. Sample (a) was fabricated at Toyota Central R&D Labs, Inc., and sample (b) at Kyoto University.
- All-solid-state LIB: A mixture of LCO, LNO, or LTO with a solid-state electrolyte $Li_7La_3Zr_2O_{12}$ (LLZ) or $Li_{10}GeP_2S_{12}$ (sulfur solid-state electrolyte). These solid-state LIB samples were fabricated at Toyota Battery Research Division, Jülich Research Center, and the Tokyo Institute of Technology for the experiments described in the following sections.

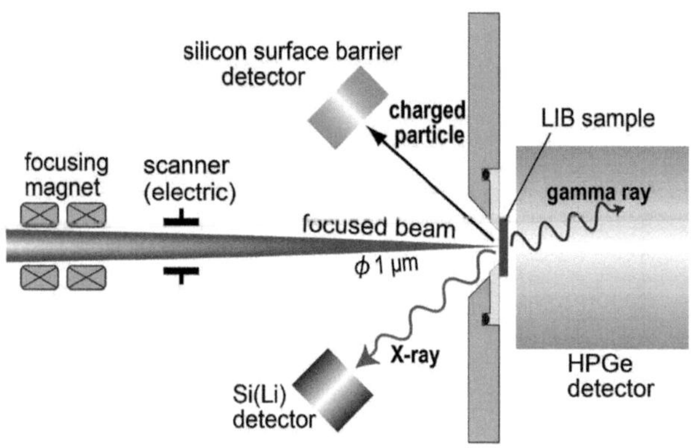

Figure 5.6 Proton microbeam diagnostic system at TIARA. The LIB sample is positioned on the boundary between the vacuum and the air regions. The ion beam irradiates a sample from the vacuum side. X-ray and charged particle detectors are located in vacuum.

The lithium concentration distribution was characterized by micro-PIGE and micro-PIXE [5]. The proton beam current was about 300 pA. The total charge of the injected proton beam for each measurement was around 0.48 μC. The beam current was measured in a conductor foil located at the sample holder. The beam diameter was 1.5 μm, and the total scan area was approximately 200 × 200 μm. The scanning step was 128 in one direction, to provide 128 × 128 space points for each image. Measurements were carried out in vacuum to achieve high beam spatial resolution. Nevertheless, because of the microbeam halo effects and the lateral straggling of the beam in penetrating into the samples, the lateral resolution of the microbeam scan was larger than the beam diameter. The γ-ray and X-ray from the sample were detected by HPGe at 0° and Si(Li) at 140°, respectively. The charged particles were also measured at 140°, as shown in Fig. 5.6. The typical measuring time was around 30 min. for taking one image. We have measured the lithium distribution of the same sample twice successively in order to confirm that the beam-induced lithium diffusion in the sample is not effective in the present experimental conditions.

5.1.3.2 Diagnostics of Li concentration and micrometer-scale imaging of Li-ion battery electrodes

Figure 5.7 shows the γ-ray and X-ray spectra of a LIB cathode irradiated by a 3 MeV proton beam. The intense γ-ray signals observed in Fig. 5.7a are ascribed to F, Li, Mg, and Al, as shown in Table 5.1. The PIGE spectral intensities of Li for a charged sample and an uncharged sample are compared in Fig. 5.7b, which clearly shows that about 10% of the Li ions is extracted in the charged positive electrode. Figure 5.7c shows K_α and K_β X-ray (PIXE) spectra of Ni and Co for the same samples of Fig. 5.7a and 5.7b. As expected, the PIXE spectra of Ni and Co are the same for the charged and uncharged samples.

Table 5.1 Energies and nuclear reactions of the γ-rays observed in the PIGE spectrum shown in Fig. 5.7a

E_γ (keV)	Reaction
110	$^{19}F(p, p'\gamma_{1\text{-}0})^{19}F$
197	$^{19}F(p, p'\gamma_{2\text{-}0})^{19}F$
429	$^{7}Li(p, n\gamma_{1\text{-}0})^{7}Be$
478	$^{7}Li(p, p'\gamma_{1\text{-}0})^{7}Li$
585	$^{25}Mg(p, p'\gamma_{1\text{-}0})^{25}Mg$
843	$^{27}Al(p, p'\gamma_{1\text{-}0})^{27}Al$
1013	$^{27}Al(p, p'\gamma_{2\text{-}0})^{27}Al$
1236	$^{19}F(p, p'\gamma_{3\text{-}1})^{19}F$
1349	$^{19}F(p, p'\gamma_{4\text{-}1+5\text{-}2})^{19}F$
1368	$^{27}Al(p, \alpha\gamma_{1\text{-}0})^{24}Mg$

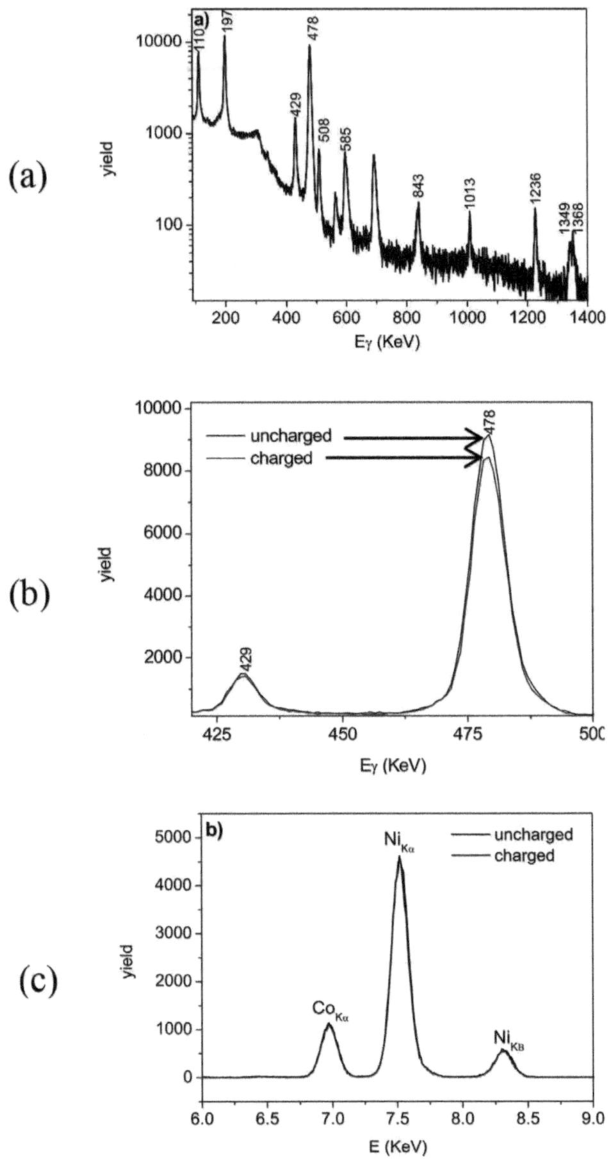

Figure 5.7 PIGE and PIXE spectra of LIB cathode. (a) γ-ray spectrum, (b) γ-ray spectra of Li, for uncharged (thin line) and charged (thick line) cathode, and (c) X-ray spectra of Ni and Co for these cases.

Figure 5.8 shows the distributions of Li and Ni obtained by PIGE and PIXE in a secondary particle with a diameter of about 10 μm. The Li 478 keV γ-ray and Ni 7.5 keV K_α X-ray were used for the imaging. The PIGE and PIXE images are compared with the SEM image of the same particle. The comparison shows that the spatial resolution of the ion beam analysis with TIARA's proton microbeam is as good as a few microns. The distribution of the secondary particles in a composite cathode was also measured, as shown in Fig. 5.9. Here, the Li and Ni distributions in the composite electrode agree well with the distribution of the secondary particles in the SEM image.

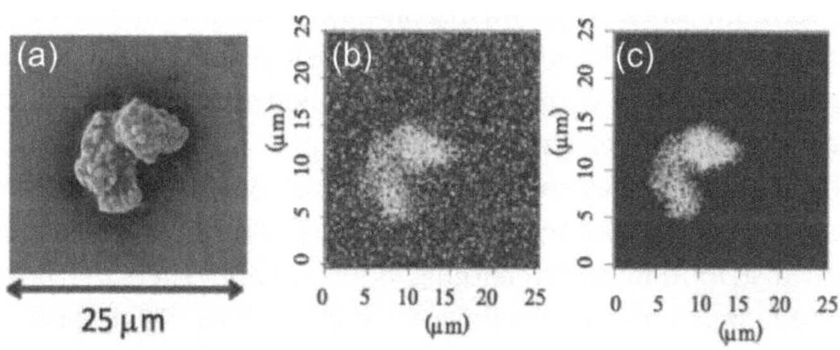

Figure 5.8 The images of an LNO secondary particle; (a) SEM image, (b) PIGE image of Li, and (c) PIXE image of Ni.

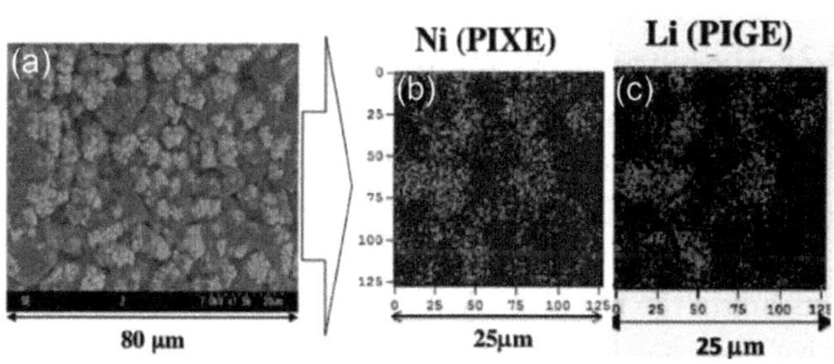

Figure 5.9 Images of an as-received LNO composite cathode; (a) SEM image, (b) PIXE image of Ni, and (c) PIGE image of Li.

A composite cathode is a mixture of secondary particles composed of active materials like LNO, LFP, LCO, carbon black, and glue coated on a metal electrode. At the interface of the secondary particles with the electrolyte, the lithium ions intercalate (during charge) or deintercalate (during discharge) through a surface electrochemical reaction. In these processes, the electrons also move in (during charge) or move out (during discharge). Therefore, not only the lithium transfer between the secondary particles and the electrolyte, but also the electrical contact of the carbon black with the secondary particles is important for LIB performance.

5.1.4 Thickness and Charge Rate Dependencies of Lithium Distributions in Charged Electrodes

5.1.4.1 Introduction

In this section is described how to measure the dependencies of the lithium distribution in the electrodes on the thickness of the electrode and the charge and discharge conditions.

For fabricating charged samples, a positive electrode is attached to a separator filled with a liquid electrolyte and a lithium metal as a negative electrode. Then, current is applied. After charging, the battery is quickly dissembled, typically within 3 min., and the liquid electrolyte is dried out to quench the Li diffusion. In this way lithium depth distribution after charging is preserved. Then the positive electrode is cut to observe its cross section. This drying process is intended not to substantially influence the lithium distribution in the electrode. Thus, the lithium distribution in the electrolyte prior to and after the drying process is assumed to be the same.

5.1.4.2 Thickness dependence of lithium distribution

To characterize the lithium battery performance, two kinds of irradiation experiments have been carried out. In the first case, two Li-ion battery positive electrodes with different thicknesses (35 μm and 105 μm) were measured to observe the electrode thickness dependence of the lithium distribution. Here, the electrodes consist of active material $Li_xNi_{0.8}Co_{0.15}Al_{0.05}O_2$ ($0.75 \leq x \leq 1$) secondary particles mixed with artificial carbon and binder, where the weight percentages are 85% for active material, 10% for carbon, and 5% for binder. This positive electrode is coated on an Al current collector of 20 μm thickness.

The results for the 105 μm and 35 μm thick electrodes are shown in Figs. 5.10a and 5.10b, respectively. The separator is on the upper side and the Al current collector on the lower side in these figures. The uniform color distribution of the Ni signals in both data shows that Ni is homogeneously distributed in these electrodes.

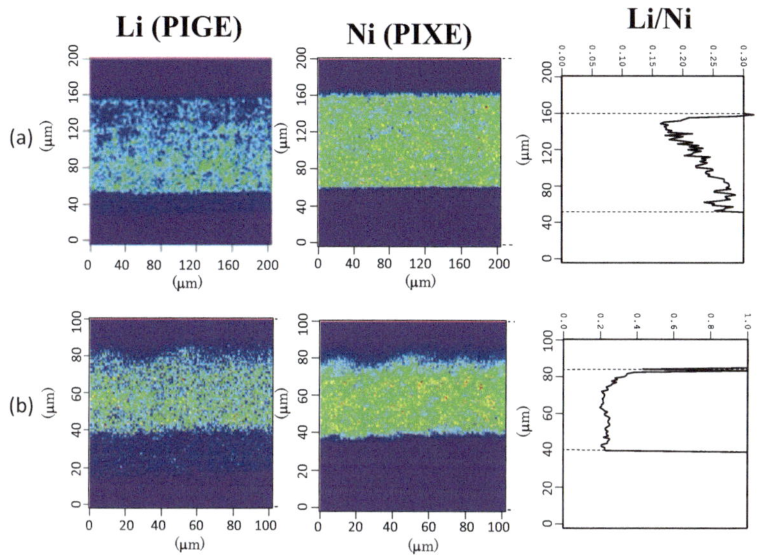

Figure 5.10 Li and Ni distributions and the Li:Ni ratio evaluated from these distributions for two positive electrodes with different thicknesses both charged at the same rate: (a) 105 μm thick sample charged for 15 min. at 2 C rate and (b) 35 μm thick sample charged in the same conditions. The dimensions of the images are (a) $200 \times 200\ \mu m^2$ and (b) $100 \times 100\ \mu m^2$.

On the other hand, an inhomogeneous lithium distribution is observed for the 105 μm thick electrode, shown in Fig. 5.10a. In particular, the lithium content close to the Al current collector of within ≤30 μm is higher than that at larger distances. This result is quantitatively presented by the Li:Ni ratio, which was calculated by integrating the Li and Ni signals for the same depth along the direction parallel to the electrode surface and then dividing the obtained values. The Li/Ni plot in Fig. 5.10a shows a monotonous increase in the Li:Ni ratio from the separator side to the Al side. In contrast, a very different situation is observed for the electrode with a thickness of 35 μm, shown in Fig. 5.10b, where the lithium distribution is almost constant along the whole thickness. These results point out that lithium is distributed more homogeneously in the thin electrode than in the thick electrode, indicating a possibility of better battery performance if thin electrodes are selected.

5.1.4.3 Charge rate dependence of lithium distribution

As the second example, the dependence of lithium distribution on the charge rate was measured. According to the literature [7], when lithium atoms are deintercalated from $Li_xNi_{0.8}Co_{0.15}Al_{0.05}O_2$ during the charge process, the lithium concentration in the positive electrode is expected to become inhomogeneous. This inhomogeneity depends not only on the electrode thickness but also on the charge rate, the electrode elemental composition, and its structure. Here we have studied the influence of the charge rate on the lithium distribution by measuring the Li and Ni distributions for the two electrodes, both with the same thickness (105 μm) but charged under different conditions—6 mA/cm^2 and 0.6 mA/cm^2. The charge parameters (current density and time) are selected in order to obtain the same charge state in both samples. The Li and Ni distributions are shown in Fig. 5.11a for the fast charge and Fig. 5.11b for the slow charge. The Ni PIXE images in Figs. 5.11a and 5.11b indicate that Ni is homogeneously distributed in the electrodes. On the other hand, the lithium images of (a) and (b) reveal that the Li-ion distribution is inhomogeneous for both electrodes. Moreover, significant differences are observed in the lithium distributions between the high charge rate and the low charge rate, as shown clearly by the Li:Ni ratio in Fig. 5.11.

For the high charge rate sample, a homogenous gradient in the lithium distribution along the depth is observed. In particular the lithium distribution linearly decreases as the distance is increased from the Al current collector. However, for the sample charged with lower current (0.6 mA/cm^2) and longer duration (2.5 h), two well-defined regions with almost constant lithium content in each but with a clear difference in the lithium content between them is observed. Quite an abrupt difference of the lithium content is observed between one region and the other. It is a new finding from the performance point of view that the lithium distribution is nonuniform even for the slow charge sample.

5.1.4.4 Summary for applicability of PIGE/PIXE to lithium-ion battery diagnostics

Micro-PIGE and micro-PIXE spectroscopy techniques have been successfully applied to accurately measure the lithium distribution in LIBs. The proton microbeam enables us to characterize the lithium distribution with a spatial (lateral) resolution in the micrometer range.

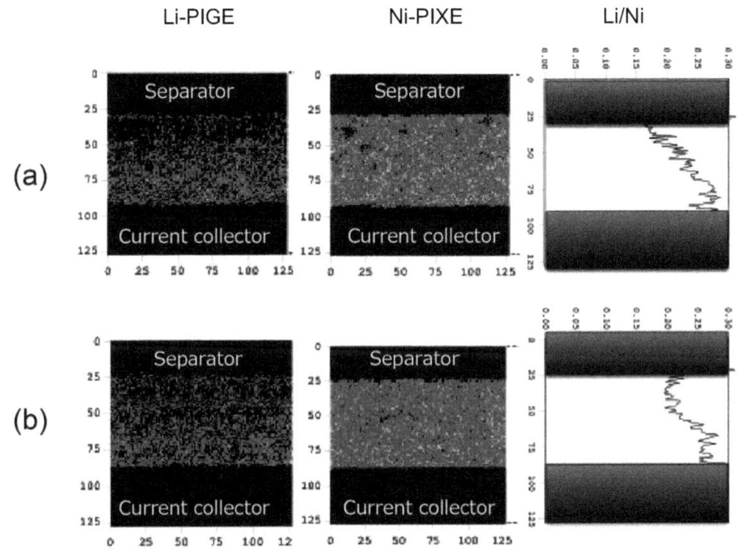

Figure 5.11 The lithium distribution for two positive electrodes of the same thickness but charged under different current densities: (a) 2 C, 15 min. and (b) 0.2 C, 150 min. The lithium content near the current collector is around 10% higher than that in the region close to the separator. Reprinted from Ref. [5], Copyright (2012), with permission from Elsevier.

The micro-PIGE images show that the lithium ions are distributed inhomogenously in the positive electrode due to a random distribution of the secondary particles, as also observed in the SEM images.

It was demonstrated that lithium distribution in the cross sections of electrodes depends on the electrode thickness and the charging conditions. The lithium distribution is homogeneous in a thin electrode, whereas it becomes very inhomogeneous in a thick electrode. For the thick electrode, a slow charge rate gives rise to a smaller gradient of the lithium distribution in the region close to the current collector, indicating that the slow charge rate is favorable from the performance point of view, since the electrochemical reaction takes place uniformly and more energy can be stored.

Micro-PIGE is a powerful tool for the diagnostics of LIBs. These results are relevant for further development of advanced LIBs with improved performances.

5.1.5 Relaxation of Li-Ion Distribution in Active Materials of Li-Ion Batteries

In the previous section, we have described that the distribution of lithium concentration over the cross section of LIB electrodes of $LiNi_{0.80}Co_{0.15}Al_{0.05}O_2$ (LNO) can be measured by an ion beam analysis with micro-PIGE and micro-PIXE. It has been found that the lithium distribution is nonuniform for thick and fast charge electrodes [5]. The electrode material dependence of the nonuniformities of the lithium distribution has been also measured by ion beam analysis. In this section, we show how to diagnose the relaxation of the lithium distribution after discharge or charge.

The relaxation of the nonuniformity of lithium concentration after charging has been investigated for fast charge thick positive electrodes made by two typical active materials: LNO and LFP. It is expected that the relaxations of lithium depth distribution for LNO and LFP are very different. Here LNO is a typical LIB material in which the lithium concentration changes continuously. The electrochemical potential of an LNO particle depends on the lithium concentration. On the other hand, LFP is another typical LIB material in which the battery reaction occurs through "the two-phase reaction." That means, LFP and $FePO_2$ phases coexist and the phase boundary propagates in the charge and discharge processes. Therefore, it is expected that the electrochemical potential of the active particle does not depend on the lithium concentration and the nonuniformity of the Li distribution in the LFP electrode does not relax when the external current is switched off.

The difference between the relaxation properties of LNO and LFP was measured as follows. The positive electrode was fabricated from the mixture of an active material (LNO or LFP), carbon black as a conductor, and a binder (PVdF) coated on a 25 μm thick Al. The ratio of these materials is 85:10:5 in weight, and the average particle diameter of the active materials is 5 ≈ 10 μm. The thickness of the positive electrode is 100 μm for LNO and 50 μm for LFP, with the areal density of about 21 mg/cm^2 for LNO and 10 mg/cm^2 for

LFP. The positive electrode with 2 cm^2 area and a lithium metal as a negative electrode were used for charging. The electrolyte was a mixture of ethyl carbonate, diethyl carbonate, and ethyl-methyl carbonate. The current densities for charging were 6 mA/cm^2 for LNO and 45 mA/cm^2 for LFP. To charge sufficiently (70 mAh/g for LNO and 105 mAh/g for LFP), these electrodes were charged for 15 min. for LNO and 1.5 min. for LFP. The average compositions of the positive material after charging are $Li_{0.75}Ni_{0.80}Co_{0.15}Al_{0.05}O_2$ and $Li_{0.5}FePO_4$.

Figure 5.12 shows a microscope image of the cross section of a positive electrode made of an active material, LNO, which was used in this study. The square area indicated by the white line in Fig. 5.12 was scanned by the proton microbeam to observe the lithium distribution relaxation. Here, the white particles are secondary particles of the active material, which are 10 μm in diameter on average, and the black particles are a mixture of conductive carbon (carbon black) and a binder (PVdF) impregnated by an electrolyte (in other words, this part corresponds to the porosity or cavity of the porous electrode). The sample of LFP also has a structure similar to the one shown in Fig. 5.12.

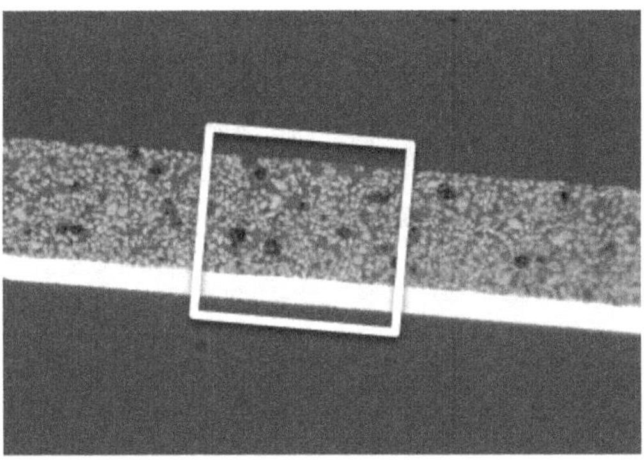

Figure 5.12 Cross section of an LNO positive electrode.

Figure 5.13a shows a variation of the charging voltage of the LIB of an LNO electrode during charge. The rapid voltage change at the beginning of the charge in Fig. 5.13a is due to the polarization of electrodes. The voltage variation is larger for a high charge rate than that for a low charge rate. The 2 C rate curve in Fig. 5.13a shows that the voltage increases initially and approaches the constant since the polarization becomes uniform within 20 min. The electrode was taken out from the cell and washed with diethyl carbonate within 3 min. after the end of charge for making a nonrelaxation sample. To investigate the relaxation, two different samples were made, where one was taken out at 1 h and the other at 10 h after the charge.

In Fig. 5.13b, the terminal voltage variation of a LIB of an LFP electrode during and after the charge is shown. To measure the lithium concentration distribution of the cross section of the electrodes, the electrodes were cut and polished by using an argon beam to make a smooth surface. LFP samples for relaxation measurements were fabricated basically in the same way as described above.

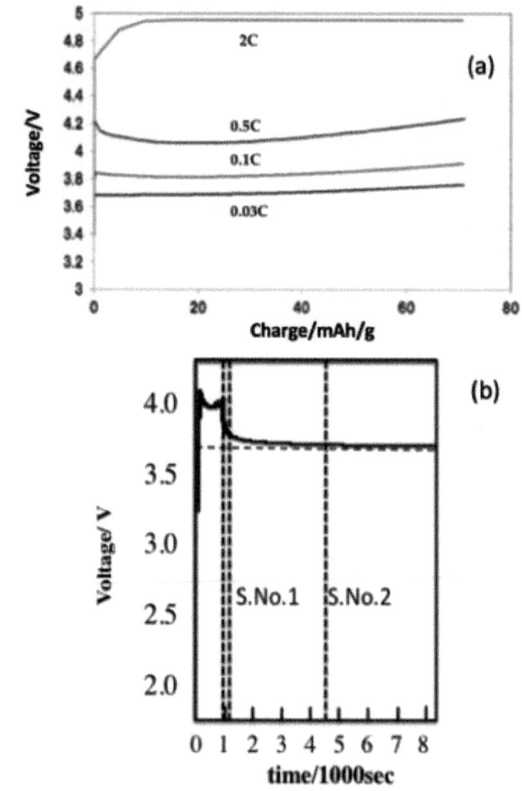

Figure 5.13 (a) Voltage vs. charging time of a LIB of an LNO positive electrode for 0.03 C, 0.1 C, 0.5 C, and 2 C charge rates. (b) Terminal voltage variation of the LFP electrode during the charge with 0.5 C rate and after the charge. The broken lines show times when the samples were taken out.

The nonuniformity of lithium distribution is expected to relax when the electrochemical potential of a secondary particle, like that of LNO, relative to the electrolyte in the pore electrode depends upon the lithium concentration of the active particle. That is, the electrochemical potential of active particles near the current collector is higher than that of active particles near the separator. Because of this potential difference, lithium ions move from active particles in the neighborhood of the current collector to those in the neighborhood of the separator. However, if there is no lithium concentration dependence of the electrochemical potential of a secondary particle, like that of LFP, there is no force for driving lithium-ion flow. Therefore, it is expected that the relaxation of distribution does not occur. Although the above difference of relaxation between an LNO electrode and an LFP electrode has been widely recognized, relaxation of the nonuniformity of the lithium-ion concentration in LIB electrodes has not been observed directly on the mesoscopic scale.

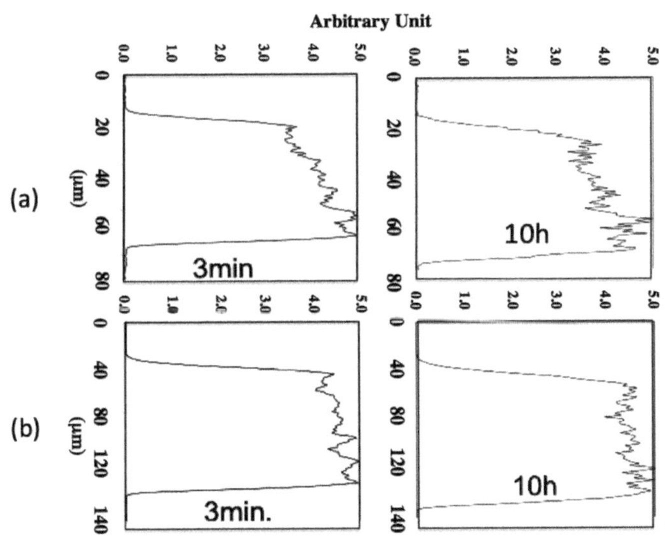

Figure 5.14 Relaxation of lithium distribution nonuniformity of $LiFePO_4$ (a) and $LiNi_{0.8}Co_{0.15}Al_{0.05}O_2$ (b) from 3 min. to 10 h after charge.

Ion beam analysis have been applied for the fast charge electrodes made by LNO and LFP to observe the relaxation of the lithium concentration nonuniformity. The electrolyte of a charged electrode is washed out to quench the battery reaction after the relaxation periods. The relaxation periods are selected to be 3 min., 60 min., and 600 min. for the present experiments.

As shown in Fig. 5.14b, the nonuniformity of lithium distributions disappears for the LNO electrode within 1 h. On the other hand, the nonuniformity remains for 10 h in the LFP electrodes. This indicates that the potential distribution inside the LFP electrode is quite uniform even for a nonuniform lithium distribution when the external current is switched off.

5.1.6 Application of Micro-PIGE and Micro-PIXE in Diagnostics of All-Solid-State Lithium-Ion Batteries

The all-solid-state LIB (hereafter referred to as ASSLIB) is one of the advanced LIBs due to its potential for high-energy-density batteries. However, Li-ion conductivity in a solid-state electrolyte and the solid-solid surface electrochemical reaction rate are not sufficient. Various aspects, such as new materials and new combination of materials, have been explored for the development of advanced ASSLIB. In this exploration, characterization of the Li-ion behavior in ASSLIB is necessary and the development of diagnostic techniques is very important.

The application of ion beam analysis to ASSLIB diagnostics will be useful for solving the above issues. Recently, lithium distribution in ASSLIB electrodes has been observed by micro-PIGE and micro-PIXE. An example of the sample design in this measurement is shown in Fig. 5.15. The all-solid-state electrolyte is LLZ ($Li_{6.6}La_3Zr_{1.6}Ti_{0.4}O_{12}$). The cathode is made of a mixture of LCO and LLZ. The test cell shown in Fig. 5.15 was charged during the measurements by extracting lithium from LCO. Since the lithium concentration per electron is lower in the LCO cathode than in the LLZ electrolyte, the PIGE signal for lithium is lower in the LCO layer, as shown in Fig. 5.16. The proton beam is scanned over the rectangular area in Fig. 5.15.

A 2D micro-PIXE image of cobalt is shown in Fig. 5.16a. Note that the interface between LLZ and LCO+LLZ is located around pixel number 40 of the vertical axis (shown by a horizontal line in Fig. 5.16b and 5.16c). Figures 5.16b and 5.16c show the line out signal intensity, integrated over the horizontal direction, of Co-PIXE and Li-PIGE, respectively. Figure 5.16c indicates that the lithium concentration between pixels 40 and 60 is lower by approximately

10% than that between pixels 10 and 30. This may be due to the lithium nonuniformity induced in the charging process. Specifically, lithium concentration near the interface is lower in comparison to that in the bulk since the electron chemical reaction is faster near the interface. The above ion beam analysis for the ASSLIB is applicable for the in-site measurement, as will be discussed in Chapter 8 [8].

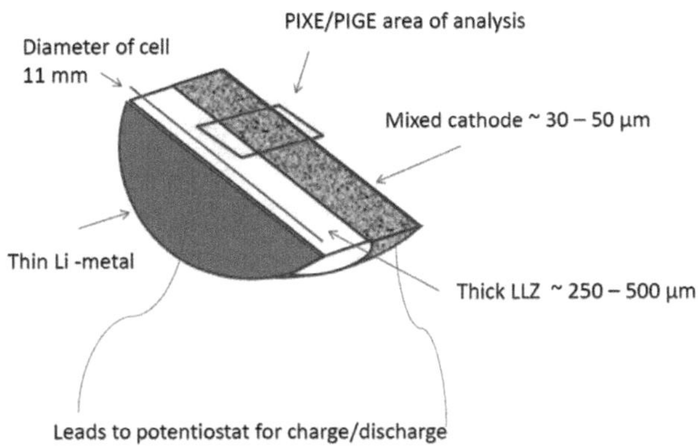

Figure 5.15 Cross section of an ASSLIB cell. The cathode is 30–50 μm thick layer of a mixture of LLZ + LCO, the electrolyte is 200–500 μm thick LLZ, and the lithium metal layer is for the anode.

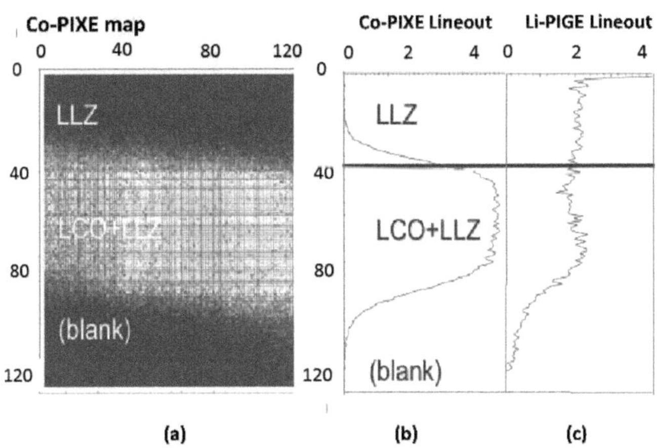

Figure 5.16 Micro–PIGE and micro-PIXE for an ASSLIB with solid-state electrolyte LLZ and the active material LCO. (a) 2D image of Co-PIXE where the scales are pixel numbers of the beam scan, (b) line out profile for Co-PIXE signal, and (c) line out profile of Li-PIGE across the cathode/electrolyte interface. The scales in (b) and (c) are the signal intensities in arbitral units.

5.2 Application of Nuclear Reaction Analysis to Lithium-Ion Battery Diagnostics

5.2.1 Introduction

As described in Section 5.2, micro-PIGE and micro-PIXE are powerful tools for characterizing elemental distribution in condensed matters nondestructively and quantitatively. NRA is another ion beam analysis method based on detecting nuclear reaction products and measuring their energy spectra. NRA is less frequently used because the cross sections for nuclear reactions are generally small, but some reactions have large cross sections (a few barns) when the projectile (p, d, α, etc.) energies are high enough (several mega-electron volts). With appropriate combinations of projectile and sample materials, it is possible to determine depth distributions of particular elements in the sample. In this section, we describe the principle of NRA and its application in characterizing the LIB. In particular, the applications of NRA in characterizing lithium depth distributions, ion beam irradiation experiments of LIB electrodes, and the methodology of data analysis are described.

5.2.2 Lithium Depth Profiling in a Lithium-Ion Battery Electrode by a $^7Li(p,\alpha)^4He$ Reaction

There are many possible nuclear reactions for profiling the lithium depth, such as $^7Li(^3He,d)^8Be$ and $^7Li(^3He,\alpha)^6Li$, $^7Li(p,\alpha)^4He$ [6, 9–14]. A comprehensive description of analyzing lithium in a material by nuclear reactions can be found in the review article by Räisänen [15] and papers cited therein. Among various nuclear reactions, the $^7Li(p,\alpha)^4He$ pick-up reaction is most useful for detecting lithium in a material as proposed by Heck [16] and Sagara et al. [17].

Figure 5.17 shows the excitation function (incident energy and angular dependence of the reaction cross section, or reaction probability) of $^7Li(p,\alpha)^4He$ [18]. θ_{LAB} is the angle of α-emission relative to the proton beam injection direction in the laboratory system. The excitation function is the maximum at an energy of around 2.9 MeV. Note that there is no sharp peak since $^7Li(p,\alpha)^4He$

is a nonresonant reaction. The released energy (Q value) by this nuclear reaction is 17.346 MeV; this energy is extremely large. As the reaction products, two α-particles are emitted in almost opposite directions to each other with energies of about 8 MeV.

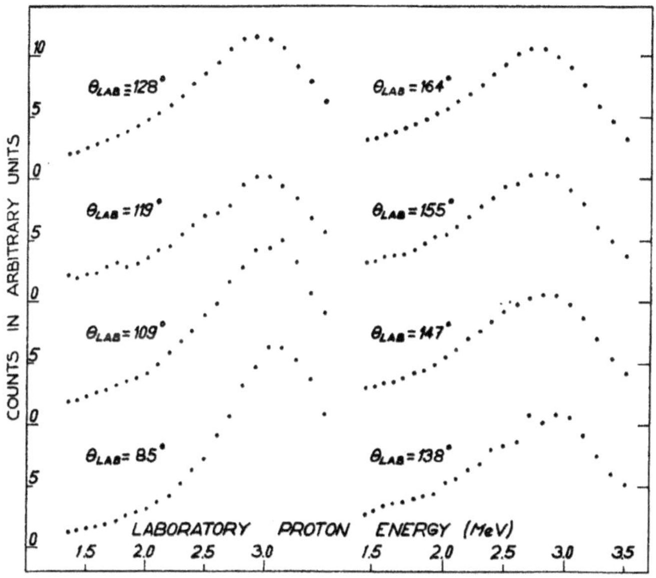

Figure 5.17 Excitation function for the ^{7}Li(p,α)^{4}He reaction. The horizontal axis is the incident proton energy, and the vertical axis shows the energy-dependent cross sections for given α-particle emission angles in the laboratory system [18]. Reprinted from Ref. [18], Copyright (1962), with permission from Elsevier.

Energy-dependent stopping lengths of 8 MeV α-particles in various materials are shown in Table 5.2. For example, the stopping length is about 43 μm in aluminum [19]. This indicates that when an α-particle is produced by the nuclear reaction in a typical LIB positive electrode (the density is about 1.5 times higher than that of aluminum) at a depth less than 20 μm, the α-particle will come out of the electrode and can be detected by charged particle detectors. The energy of the α-particle decreases when the depth of the emission point increases. Therefore, the energy spectrum of the α-particles is related to the depth distribution of lithium. The spatial resolution of this depth profile measurement depends upon several factors, such as the energy resolution of the detector, the α-particle struggling in the sample, and proton beam energy spread. In particular, it is necessary to use an energy-sensitive charged particle detector.

Table 5.2 Examples of the stopping lengths of 8 MeV helium in materials (calculated by SRIM2008)

Material	Stopping length (8 MeV ^{4}He) [μm]
PET	58.82
Aluminum	42.99
Silicon	49.21
Iron	19.05
Copper	19.30
Gold	15.85

5.2.3 Resonant Nuclear Reaction Analysis for Lithium Depth Profiling by ^{7}Li(p,γ)^{8}Be

Another candidate for detecting lithium in a material is the ^{7}Li(p,γ)^{8}Be proton capture reaction [13, 20]. This reaction also has a characteristic large Q value of 17.254 MeV. In addition, this reaction is a resonant nuclear reaction and the reaction cross section has a strong dependence on proton incident energy. It has several discrete and sharp peaks called resonance in the low-incident-energy region. This feature has a close relation to the energy levels of the resultant nuclei ^{8}Be. The lowest resonance is located at 0.441 MeV proton incident energy [21]. At this energy, the excitation function is maximum with a very narrow resonance width of 12.5 keV in full width at half maximum.

Using this characteristic, we can obtain the depth profile of the lithium density in material. Let us consider irradiating a material containing lithium atoms with protons. We note that the proton energy decreases gradually during penetration into the material. When the incident proton energy is 0.441 MeV, it can be captured by the ^{7}Li nucleus on the surface of the material. When the incident energy is higher than 0.441 MeV, the proton gradually decreases in energy during propagation in the material until the proton energy becomes 0.441 MeV. Then, the proton is captured by ^{7}Li nucleus to emit ^{8}Be. Since not only the proton but also all the ions propagate along almost a straight line in a material, the penetration length is almost the same as the distance from the surface if the protons are irradiated perpendicularly to the surface of the sample. This characteristic indicates that we can detect lithium atoms in material at a certain depth. The accuracy of the depth profiling by resonant NRA corresponds theoretically to the distance at which a 0.441 MeV proton loses its energy by 12.5 keV (the width of the resonance), which is about 0.1 μm.

The nuclear reaction $^{7}Li + p \rightarrow {}^{8}Be^{*} \rightarrow {}^{8}Be + \gamma$ is a nondirect nuclear reaction that has the intermediate state $^{8}Be^{*}$ (an excited state of ^{8}Be). When an incident proton collides with ^{7}Li, the proton is captured and a compound nucleus $^{8}Be^{*}$ is created as an intermediate state of the nuclear reaction. This $^{8}Be^{*}$ is an excited state nucleus and de-excites immediately, in most cases, emitting γ-ray(s). When a proton with the energy of 0.441 MeV is captured by a ^{7}Li nucleus, an excited state of ^{8}Be is created with an excitation energy of 17.64 MeV, which then emits γ-rays immediately. Their energies are 14.586 MeV and 17.619 MeV according to the final state of ^{8}Be after the γ emission transition.

We can find the existence of ^{7}Li nuclei by detecting these γ-rays. Both final states decay by two α-particles immediately. However, in most cases they do not come out of a sample material because their kinetic energies are not high. Because of this feature, observing lithium nuclei by detecting these α-particles is not practical. We note that the 0.441 MeV incident energy is the lowest resonance energy of the $^{7}Li(p,\gamma)^{8}Be$ reaction. Observing the γ-rays by increasing the incident energy from this resonance energy corresponds to observing lithium in a material by increasing the distance from the surface, thus enabling the depth profiling of lithium in a material.

However, this method is valid only when it has only one resonance, in this case, 0.441 MeV incident energy. The second resonance appears in the incident energy of 1.03 MeV [21]. It corresponds to the excited states with the energy of 18.15 MeV in ^{8}Be. This state also emits a high-energy γ-ray, but the energy difference between this γ-ray and the γ-ray from de-excitation of the 17.64 MeV state is so small that we can hardly distinguish these two γ-rays (18.1 MeV and 17.6 MeV). This means that from the γ-rays emitted by a

$^{7}Li(p, \gamma)^{8}Be$ reaction with an incident energy of more than 1.03 MeV, it cannot be determined whether it comes from the inside (0.441 MeV resonance) or from the surface (1.03 MeV resonance) of the material. Therefore, the depth profile of lithium by observing the γ-rays is limited. The distance at which the 1.03 MeV proton loses 0.441 MeV energy in a typical LIB material with a density of 5 g/cm^3 is about 5 μm. Therefore, the $^{7}Li(p, \gamma)^{8}Be$ reaction can be only used for obtaining a depth profile of lithium within several micrometers from the surface.

The $^{7}Li(p,n\gamma)^{7}Be$ nuclear reaction is also available for NRA for ^{7}Li in a material. Its Q value is –1.664 MeV. The lowest excited state (first excited state) of ^{7}Be is 0.429 MeV, which de-excites immediately by emitting a γ-ray of 0.429 MeV. Although we can also observe the lithium in a material using this reaction, the reaction cross section is smaller than that of $^{7}Li(p,p'\gamma)^{7}Li$ inelastic scattering. In addition, a larger incident energy is required because of its negative Q value. As a result, the $^{7}Li(p,n\gamma)^{7}Be$ reaction may not be realistic, although it might be useful when we have to observe lithium located within several micrometers from the surface.

5.2.4 Experimental Instruments for Nuclear Reaction Measurements

In NRA, as in PIXE and PIGE, we need to observe radiations and measure their energies. Among various radiations, observing neutrons is not convenient because it is difficult to detect them and to measure their energies. Therefore, in most cases, observing γ-rays and charged particles is practical for NRA use. In this section, we introduce the detectors for charged particles and high-energy γ-rays. For low- and middle-energy (about several hundred keV) γ-rays, an HPGe detector and a thallium-activated sodium iodide, NaI(Tl), detector are often used. See Section 4.1.4, Chapter 4.

In the case of NRA for lithium observation with a $^{7}Li(p,\alpha)^{4}He$ reaction, a silicon surface barrier detector is suitable for detecting the α-particles from the sample materials. This detector has typically a sensitive area of 100 mm^2 and a depletion layer thickness of several hundred micrometers. Transmission-type detectors are also available; they are often used for identifying the charged particles. We should choose a detector with a thickness of about

100 μm or more to stop the α-particles, because the penetration length of a 10 MeV α-particle in silicon is about 70 μm. However, since a thicker silicon detector may result in reduced energy resolution, the detector thickness of about 100 μm is commonly used.

The kinetic energy of the α-particle from the (p,α) reaction depends on its direction of emission. The relation between the kinetic energy of the α-particle emitted by the ^{7}Li(p,α)^{4}He reaction and the emitting angle is shown in Fig. 5.18. It clearly shows that the α energy depends on the angle strongly around 90° and weakly around 0° (forward) and 180° (backward). This indicates that when the detector is set at the backward angle we can detect the α-particles with better energy resolution. (It should be noticed that positioning the detector at the forward angle is not realistic in most cases because most samples have enough thickness for stopping the emitting α-particles.) Therefore, the α-particle detector should be positioned at the backward angle. In addition to the energy resolution, positioning the detector at the backward angle brings another advantage in analyzing the experimental data. The relation between the energy of the α-particle emitted at the backward angle of 175° and the incident energy of the proton is shown in Fig. 5.19 [16]. For the ^{7}Li(p,α)^{4}He reaction, the α-particle energy is relatively constant and is higher than those of α-particles from other (p,α) reactions with the different targets. These characteristics make the analysis of the experimental data easy and clear.

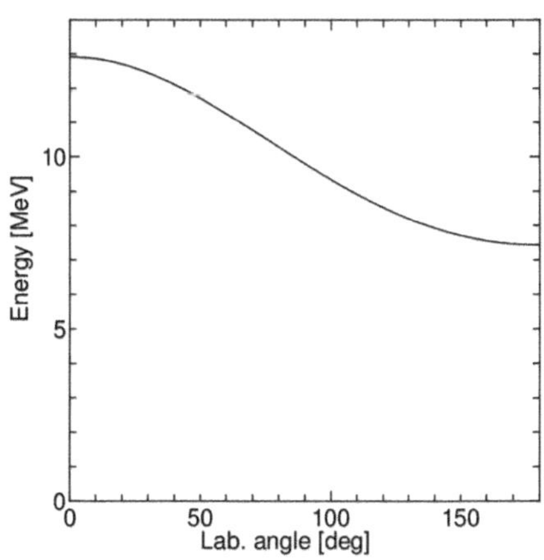

Figure 5.18 Angular dependence of the emitted α-particles from the ^{7}Li(p,α)^{4}He reaction. The incident energy of the proton is 3 MeV.

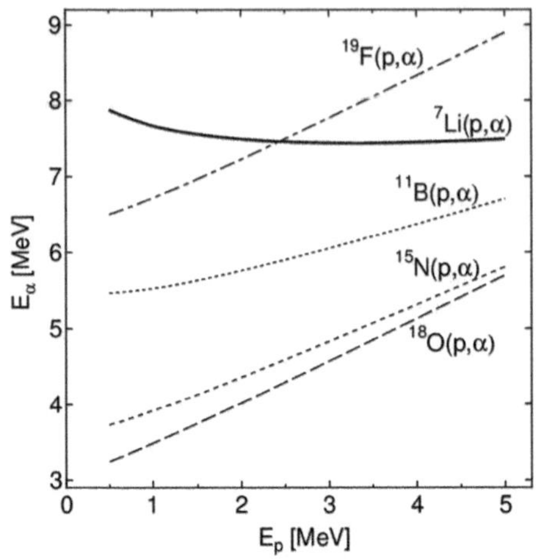

Figure 5.19 Emitted α-particle energy versus incident proton energy for the $^7Li(p,\alpha)^4He$ reaction with a high Q value and for other nuclear reactions. The emitted angle is 175° at the laboratory system [16].

For γ-ray detection, an HPGe detector is usually used because of its high energy resolution. However, HPGe is not suitable for NRA of lithium because the γ-ray energies are several mega–electron volts or more, where its detection efficiency is low. Inorganic scintillators, such as NaI(Tl) and bismuth germanate (BGO), are often used for NRA of lithium.

NaI(Tl) is the most well-known inorganic scintillator. It contains a small amount of thallium in the crystal, which works as an activator. The energy resolution is relatively high among the inorganic scintillators because of its high luminescence efficiency. On the other hand, NaI(Tl) crystal is hygroscopic and must be housed in an airtight protective enclosure. As a result, NaI(Tl) cannot be used for charged particle detection. The shape of the NaI(Tl) crystal is typically cylindrical or rectangular parallelepiped, several centimeters in length.

BGO ($Bi_4Ge_3O_{12}$) is also an inorganic scintillator that is often used for detecting high-energy γ-rays. Relative to NaI(Tl), it has better transparency and higher density (7.13 g/cm^3) and contains higher-Z elements (Bi:83), leading to higher detection efficiency for high-energy γ-rays per unit volume. Its firmness and nonhygroscopicity provide ease of handling. However, the light yield from a BGO crystal is lower and thus the energy resolution is worse than that from a NaI(Tl) scintillator. The shape and size of the BGO crystal is similar to those of NaI(Tl).

When a high-energy photon with several mega–electron volts or more energy travels in a material, it loses all of its energy by pair production or a part of it by Compton scattering. (The probability of photoelectric absorption is extremely low.) The electron created by pair production loses its kinetic energy by ionizing or exciting atoms consisting of the material and stops soon. The positron also loses its kinetic energy by ionizing or exciting atoms, but on the other hand, it combines with a normal electron (negative electron) just before stopping. When the positron-electron pair disappears, two photons with the energy of 0.511 MeV are produced. These photons have larger probabilities of photoelectric absorption than that of the initial several-mega–electron volt photon. However, because the size of the detecting material is finite, a fraction of the produced photons escapes from the material.

Therefore, there are four cases of energy deposition from the two photons to the material, as shown below:

1. Both photons deposit all of their energies to the material and disappear.
2. One photon deposits all of its energy to the material and disappears, and the other escapes from the material without any interaction.
3. Both photons escape from the material without any interaction.
4. One or both photons are scattered by an electron, losing a part of their energies, and escape from the material.

These events form (1) a full-energy peak, (2) a single-escape peak, (3) a double-escape peak, and (4) Compton continuum in the photon energy spectrum. The full-energy peak shows the energy of the photons. The single-escape peak appears at an energy of 0.511 MeV below the full-energy peak and the double-escape peak at an energy of 1.022 MeV (double of 0.511 MeV) below it.

It should be noted that the existence of natural radioactivity in the environment affects the experimental results of γ-ray measurements. Natural radioactivity exists all over the earth. The important components are potassium, thorium, uranium, and the members of the decay chains, starting from thorium and uranium.

For the measurement of medium- or high-energy γ-rays, the most important radioactive nuclei are ^{40}K and ^{208}Tl. ^{40}K has an extremely long half-life, of 1.277×10^9 years, and therefore, it makes up 0.0117% of the total amount of potassium found in nature. The half-life of ^{208}Tl is 3.053 min., but it is created continuously through the decay chain starting from ^{232}Th, whose half-life is 1.405×10^{10} years. ^{40}K emits a 1.461 MeV γ-ray just after electron capture, and ^{232}Th emits a 2.615 MeV γ-ray just after β decay. These γ-rays form full-energy peaks, single- and double-escape peaks, and Compton continuous background events in the γ-ray spectra. On the other hand, there is no prominent γ-ray higher than 2.615 MeV in natural activity. Therefore, γ-rays with energies higher than 2.615 MeV are desirable for observing NRA.

5.2.5 Simulation for Nuclear Reaction Analysis

In general, it is not feasible to analytically obtain nuclear reaction probabilities (cross sections) because the mechanisms of nuclear reactions are complicated and many properties of nuclei and nuclear interactions are not well known. However, we can predict the energy spectra of the charged particles emitted by nuclear reactions using the reaction cross sections obtained from former experiments. By comparing the experimental spectrum with the simulation, we can estimate the depth profiles of the sample materials.

The code SIMNRA, developed by Matej Mayer, is a commercial program for simulation of the spectra by ion scattering. SIMNRA can simulate energy spectra of the ions not only elastically or inelastically scattered by the target nuclei but also emitted by nuclear reactions. We can select an incident particle from ^{1}H(p), ^{2}H(d), ^{3}H(t), ^{3}He, ^{4}He(α), and other heavy ions. The maximum number of layers of the target in simulation is 100, allowing us to express realistic structures of the target, such as a packing film or an oxidation and contamination layer on the surface. Each layer consists of multiple elements distributed homogeneously, and the ratio of the elements can be defined arbitrarily. In addition, we can also simulate energy spectra with a setting where thin films are placed in front of the charged particle detector for protection from irradiated heavier particles.

SIMNRA has various experimental data, including the cross sections of non-Rutherford scattering and nuclear reactions. Therefore, the energy spectra of common nuclear reactions, such as (p,α) reaction, can be obtained. Moreover, the user can add experimental cross-section data if necessary for simulating energy spectra. The user compares the simulated energy spectra with the experimental one, and the parameters (thickness of the target, ratio of the contained atoms, etc.) are adjusted to reproduce the experimental result. The parameter set that can reproduce the experimental result provides information on the characteristics of the target, such as the ratio of elements and their depth profiles.

5.2.6 Examples of Simulations for Nuclear Reaction Analysis

A comparison of the energy spectra obtained from an experiment and a simulation is described for the two cases we have studied, since this comparison process will be useful in understanding both physics and experiment in real situations.

5.2.6.1 Example A: Reproducing the Rutherford backscattering spectrum

The sample is a positive electrode–active material deposited on an aluminum film. Its chemical formula is $Li_{1-x}NiO_2$, with a small amount of cobalt, magnesium, and aluminum contained in place of nickel. Here $x = 0$ indicates the state of full discharge and x is approximately 0.5 when the sample is fully charged. The thickness of the active material and the aluminum film is 27 μm and 22 μm, respectively. Since the active material before charging is $x = 0$, its chemical formula can be expressed as $LiNi_{0.85}Co_{0.10}Al_{0.05}O_2$. This composition is constant along the depth since measurement was made on the active material without charging.

The experimental setup is shown in Fig. 5.20. The sample was irradiated by a 3 MeV proton microbeam in the area of 25 × 25 μm^2, over which the proton beam was scanned to obtain average information in this area. An annular-type silicon surface barrier detector, positioned at 175° from the beam direction, was used for detecting the charged particles from the sample material.

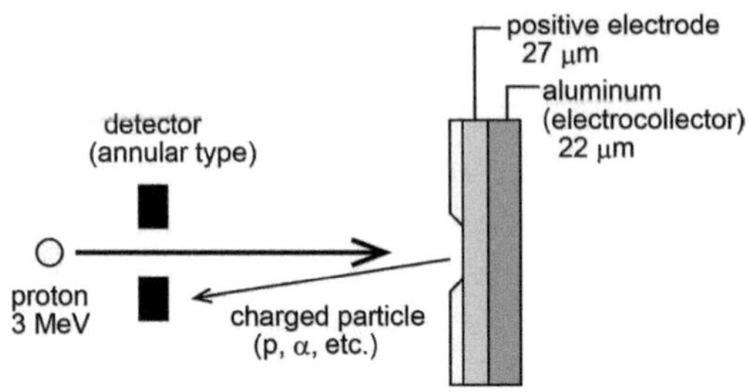

Figure 5.20 Experimental setup for observing the α-particles generated by a $^7Li(p,\alpha)^4He$ reaction. The detector is an annular-type silicon semiconductor detector located at 175° from the proton beam direction.

The experimental and simulated results by SIMNRA are shown in Fig. 5.21. In these spectra, the enormous events that can be seen in the area of less than 3000 keV are due to elastically scattered proton events. On the other hand, the events between 4300 and 7400 keV show the $^7Li(p,\alpha)^4He$ nuclear reaction events. This has information on the depth profile of the lithium atoms in the active material.

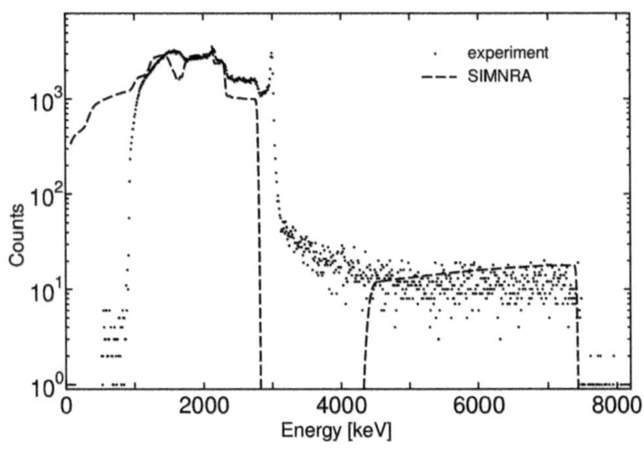

Figure 5.21 Energy spectra of the charged particles, where the experimental data are shown by solid circles and the SIMNRA simulation by a dashed line.

There are several points where the simulation cannot reproduce the experimental energy spectra.

1. There is no event in the experiment below 1000 keV.
2. A small peak at 2100 keV in the experiment cannot be reproduced.

3. The simulation between 2300 and 2800 keV is lower than the experiment.
4. A peak in the experiment at around 3000 keV cannot be predicted by the simulation.
5. A continuous region of the experiment between 3000 and 4200 keV cannot be predicted by the simulation.

These discrepancies are examined one by one.

1. There is no event in the experiment below 1000 keV.

   Low-energy events were not acquired due to a low-energy discriminator in the electronic circuit or data acquisition software. This discrepancy is not a problem because there is no information on lithium in this spectral region.
2. A small peak at 2100 keV in the experiment cannot be reproduced.

   This sharp peak is probably due to elastic scattering by carbon on the surface of the sample. It is well known that carbon atoms are deposited at the beam irradiation position on the sample when the vacuum condition is not very good. It is realistic to assume that there exists a thin carbon layer on the sample.
3. The simulation between 2300 and 2800 keV is lower than the experiment.

   This may be because the relative amounts of Ni, Co, and Al atoms to those of Li and O assumed in the simulation are small. In this case, we should increase the amounts of Ni, Co, and Al while keeping the ratio of Ni:Co:Al = 0.85:0.10:0.05.
4. A peak in the experiment at around 3000 keV cannot be predicted by the simulation.

   This peak is most probably due to protons that directly reach the silicon surface barrier detector from the behind the detector (or scattered by a beam line component). We can avoid these events by careful experimental setup.
5. A continuous region of the experiment between 3000 and 4200 keV cannot be predicted by the simulation.

   These are due to pileup of the enormous proton elastic scattering events. These pileup events overlap with ^{7}Li(p,α)^{4}He events, and therefore we have to control the pileup events by careful settings of the experimental conditions. For example, we should keep the count rate of the charged particles low and set the shaping time of the linear amplifier at a smaller value.

The result of the simulation after modifying the conditions (change the atomic ratio of the target material and add a thin carbon layer on the target) is shown in Fig. 5.22. The information of the depth profile of lithium in the sample material is contained in the region between 4300 and 7400 keV in the spectrum. The major advantage of the $^7Li(p,\alpha)^4He$ reaction as a method of observing lithium in a material is that these reaction events can be obtained separately from the enormous elastic scattering events at lower energies. In more detail, the observable depth is restricted by the condition that the elastic and reaction events are separated in the spectrum. This condition depends on the energy of the incident proton and the emission angle of the α-particles, that is, the angle of the detector position of the charged particles.

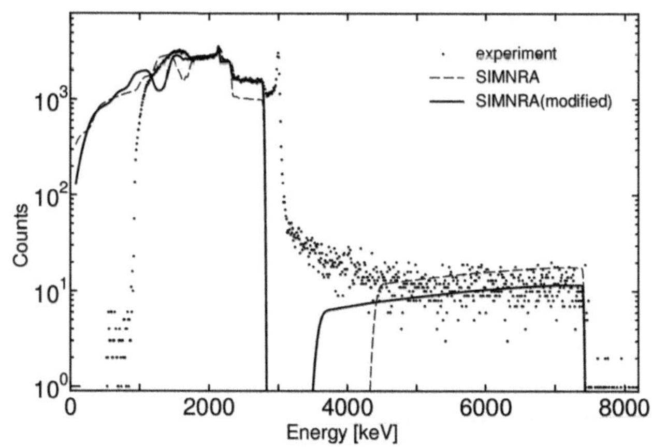

Figure 5.22 Energy spectra of the charged particles, where the experimental data are shown by solid circles and the SIMNRA simulation by a solid line. The ratio of the atoms in the positive electrode and some other conditions were modified to reproduce the experimental results. The simulated result before modification is also shown for comparison by a dashed line.

In addition, since the event rate of the elastic scattering is much higher than that of the (p,α) events, the proton scattering rate determines the upper limit of the counting rate of the charged particle detector. As a result, it needs a long accumulation time to record the $^{7}Li(p,\alpha)^{4}He$ events in order to obtain the depth profile of lithium with good statistical conditions. This means that irradiation by a large amount of protons is necessary to obtain the lithium data. Since these proton beams might cause damage to the sample, it is necessary to check beforehand using a short-period irradiation test that the proton irradiation does not affect the distribution or density of lithium in the sample.

5.2.6.2 Example B: Estimation of lithium depth distribution in a lithium-ion battery

NRA was applied to estimate the distribution of the lithium ions in the charged electrode of a lithium-ion battery [6]. As an example, the energy spectrum of the charged particle is shown for the positive electrode material of LFP in Fig. 5.23. It shows spectra of RBS protons and α-particles due to $^{7}Li(p,\alpha)^{4}He$ and $^{19}F(p,\alpha)^{16}O$ reactions. The LFP positive electrode used in this measurement was fabricated at the Uchimoto Laboratory of Kyoto University. This sample is a mixture of the active particles of LFP with artificial carbon (electric conductor) and binder. The active particles 0.5–1 μm in diameter are randomly mixed with the carbon and the binder. The thickness and the areal density of the positive electrode are 30 μm and 1.0 mg/cm^2, respectively. The porosity of the electrodes is 53%. The LFP active material is coated on an aluminum foil of 25 μm thickness. This electrode was charged and the lithium ions were deintercalated from LFP. The current density during charging was 45 mA/cm^2, and the charging time was 1 min. Of the lithium, 25% was extracted by the charge. Finally, the electrode was disassembled and washed to remove the electrolyte.

We have characterized the lithium depth of the above LFP sample at the standard beam line of the National Accelerator Center, at the University of Seville [22]. In this measurement, a proton beam irradiated the sample surface at normal incidence at an energy of 3 MeV. The beam spot size and the current were 2×2 mm^2 and ~15 nA, respectively. α-particles produced in the nuclear reaction and the backscattered protons from Fe, P, O, and so on were simultaneously detected by a silicon surface barrier detector with an active area of 50 mm^2, located 10 cm from the sample at an angle of 150°.

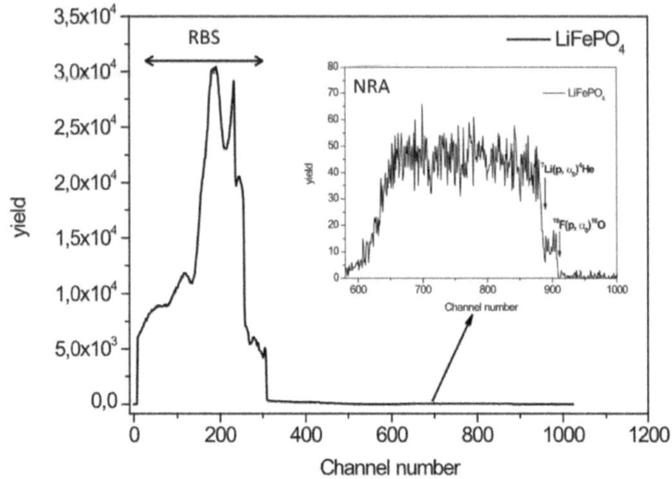

Figure 5.23 An example of a charged particle energy spectrum for the NRA experiment of the LIB positive electrode: $LiFePO_4$. Reprinted from Ref. [6], Copyright (2015), with permission from Elsevier.

From the α-particle spectrum of $^7Li(p,\alpha)^4He$, the lithium depth profile can be determined by comparing the experimental spectra with the simulation. Cross-section inputs related to the lithium and fluorine nuclear reactions were taken from Paneta et al. [23] and introduced as a R33 file in the SIMNRA code.

First the α-particle spectrum was calculated by assuming that lithium ions are distributed homogeneously in the electrode. As shown by the red curve in Fig. 5.24a, the measured and calculated spectra are clearly different. The measured α-yields for the channels between 600 and 750, which correspond to the α-particles generated deep inside the sample, are significantly higher than the calculated yields. This indicates that the lithium concentration is not homogeneous along the depth and should be higher deep inside.

A better fit of the simulation spectrum with the experimental one is obtained by assuming that the lithium distribution is not homogeneous. To calculate the spectrum, the sample is divided into several layers in depth, each layer with a different lithium concentration. A good fit is achieved after many simulations by varying the lithium distributions. As shown in Figs. 5.24c and 5.24d, the best fit is obtained for the 12 layers, where the deeper layers have higher concentrations of lithium. The lithium concentration increases from 4.9% (by the number of atoms) for the layer close to the separator up to 6.9% for the layer close to the current collector. The lithium concentrations over the 12 layers are schematically shown in Fig. 5.24d, where the left-hand side was close to the separator and the right-hand side was attached to the current collector. These results show that the electrochemical reaction was faster near the separator and slower at the layer deeper from the separator. This example shows that NRA is very useful in analyzing the distribution of lithium ions in the LIB electrodes.

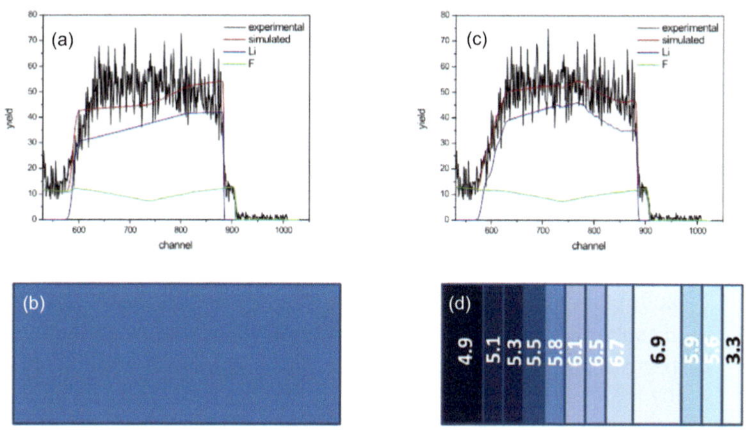

Figure 5.24 Deduction of lithium depth profile from the α-particle energy spectrum and the SIMNRA simulations. (a) The simulation results are shown by a blue line for $^{7}Li(p,\alpha)^{4}He$, a green line for $^{19}F(p,\alpha)^{16}O$, and a red line for the total when the depth distribution of lithium is assumed to be uniform, as shown in (b). (c) The best fit simulations results, where the lithium depth distribution is shown in (d). Reprinted from Ref. [6], Copyright (2015), with permission from Elsevier.

5.3 Summary

The application of PIXE and PIGE for diagnostics of LIBs is described in Section 5.1. Characterization of the depth distribution of the lithium concentration by using a proton microbeam is described in detail. Some examples of micro-PIGE/PIXE imaging of the lateral and depth distribution of lithium are described. In particular, demonstration of observations of the charge rate and electrode thickness dependencies of the lithium depth distribution is given.

In Section 5.2, the application of NRA for characterizing the lithium depth distribution is reviewed. Described are the measurement of the α-particle energy spectra produced by a proton-Li nuclear reaction and data analyses to deduce the lithium depth distribution from the α-particle spectra.

References

1. Wang, Y. and Nastasi, M. (2010). *Handbook of Modern Ion Beam Materials Analysis*, 2nd ed. (Cambridge University Press).
2. Sakai, T., Kamiya, T., Oikawa, M., Sato, T., Tanaka, A. and Ishii, K. (2000). Development of in-air micro-PIXE analysis and data sharing systems in JAERI Takasaki, *Int. J. PIXE*, **10**, pp. 91–95.
3. Tadić, T., Jakšić, M., Capiglia, C., Saito, Y. and Mustarelli, P. (2000). External microbeam PIGE study of Li and F distribution in PVdF/HFP electrolyte gel polymer for lithium battery application, *Nucl. Instrum. Methods Phys. Res., Sect. B*, **161**, pp. 614–618.
4. Tadić, T., Jakšić, M., Medunić, Z., Quartarone, E. and Mustarelli, P. (2001). Microbeam studies of gel–polymer interfaces with Li anode and spinel cathode for Li ion battery applications using PIGE and PIXE spectroscopy, *Nucl. Instrum. Methods Phys. Res., Sect. B*, **181**, pp. 404–407.
5. Mima, K., Gonzalez-Arrabal, R., Azuma, H., Yamazaki, A., Okuda, C., Ukyo, Y., Sawada, H., Fujita, K., Kato, Y., Perlado, J. M. and Nakai, S. (2012). Li distribution characterization in Li-ion batteries positive electrodes containing $Li_xNi_{0.8}Co_{0.15}Al_{0.05}O_2$ secondary particles ($0.75 \leq x \leq 1.0$), *Nucl. Instrum. Methods Phys. Res., Sect. B*, **290**, pp. 79–84.
6. Gonzalez-Arrabal, R., Panizo-Laiz, M., Fujita, K., Mima, K., Yamazaki, A., Kamiya, T., Orikasa, Y., Uchimoto, Y., Sawada, H., Okuda, C., Kato, Y. and Perlado, J. M. (2015). Meso-scale characterization of lithium distribution in lithium-ion batteries using ion beam analysis techniques, *J. Power Sources*, **299**, pp. 587–595.
7. Ogihara, N., Kawauchi, S., Okuda, C., Itou, Y., Takeuchi, Y. and Ukyo, Y. (2012). Theoretical and experimental analysis of porous electrodes for lithium-ion batteries by electrochemical impedance spectroscopy using a symmetric cell, *J. Electrochem. Soc.*, **159**, pp. A1034–A1039.
8. Yamazaki, A., Orikasa, Y., Chen, K., Uchimoto, Y., Kamiya, T., Koka, M., Satoh, T., Mima, K., Kato, Y. and Fujita, K. (2016). In-situ measurement of the lithium distribution in Li-ion batteries using micro-IBA techniques, *Nucl. Instrum. Methods Phys. Res., Sect. B*, **371**, pp. 298–302.

9. Malmberg, P. R. (1956). Elastic scattering of protons from Li^7, *Phys. Rev.*, **101**, pp. 114–118.

10. Sarma, N., Jayaraman, K. S. and Kumar, C. K. (1963). Mechanism of the Li^7 (p, α) He^4 reaction, *Nucl. Phys.*, **44**, pp. 205–211.

11. Forsyth, P. D. and Perry, R. R. (1965). The $Li^7(He^3, \alpha)Li^6$ reaction between 1.3 and 5.5 MeV, *Nucl. Phys.*, **67**, pp. 517–528.

12. Dieumegard, D., Maurel, B. and Amsel, G. (1980). Microanalysis of flourine by nuclear reactions, *Nucl. Instrum. Methods*, **168**, pp. 93–103.

13. Räisänen, J. and Lappalainen, R. (1986). Analysis of lithium using external proton beams, *Nucl. Instrum. Methods Phys. Res., Sect. B*, **15**, pp. 546–549.

14. Caciolli, A., Chiari, M., Climent-Font, A., Fernández-Jiménez, M. T., García-López, G., Lucarelli, F., Nava, S. and Zucchiatti, A. (2006). Measurements of γ-ray emission induced by protons on fluorine and lithium, *Nucl. Instrum. Methods Phys. Res., Sect. B*, **249**, pp. 98–100.

15. Räisänen, J. (1992). Analysis of lithium by ion beam methods, *Nucl. Instrum. Methods Phys. Res., Sect. B*, **66**, pp. 107–117.

16. Heck, D. (1988). Three-dimensional lithium microanalysis by the $^7Li(p, \alpha)$ reaction, *Nucl. Instrum. Methods Phys. Res., Sect. B*, **30**, pp. 486–490.

17. Sagara, A., Kamada, K. and Yamaguchi, S. (1988). Depth profiling of lithium by use of the nuclear reaction $^7Li(p, \alpha)^4He$, *Nucl. Instrum. Methods Phys. Res., Sect. B*, **34**, pp. 465–469.

18. Cassagnou, Y., Jeronymo, J. M. F., Mani, G. S., Sadeghi, A. and Forsyth, P. D. (1962). The $Li^7(p, \alpha)\alpha$ reaction, *Nucl. Phys.*, **33**, pp. 449–457.

19. Ziegler, J. F., Ziegler, M. D. and Biersack, J. P. (2010). SRIM: The stopping and range of ions in matter, *Nucl. Instrum. Methods Phys. Res., Sect. B*, **268**, pp. 1818–1823.

20. Marion, J. B. and Wilson, M. (1966). The $^7Li(p, \gamma)^8Be^*$ reaction and single-particle levels in ^8Be, *Nucl. Phys.*, **77**, pp. 129–148.

21. Ajzenberg-Selove, F. (1988). Energy levels of light nuclei A = 5–10, *Nucl. Phys. A*, **490**, pp. 1–225.

22. García López, J., Ager, F. J., Barbadillo Rank, M., Madrigal, F. J., Ontalba, M. A., Respaldiza, M. A. and Ynsa, M. D. (2000). CNA: the first accelerator-based IBA facility in Spain, *Nucl. Instrum. Methods Phys. Res., Sect. B*, **161**, pp. 1137–1142.

23. Paneta, V., Kafkarkou, A., Kokkoris, M. and Lagoyannis, A. (2012). Differential cross-section measurements for the $^7Li(p,p_0)^7Li$,

$^{7}Li(p,p_1)^{7}Li$, $^{7}Li(p,\alpha_0)^{4}He$, $^{19}F(p,p_0)^{19}F$, $^{19}F(p,\alpha_0)^{16}O$ and $^{19}F(p,\alpha_{1,2})^{16}O$ reactions, *Nucl. Instrum. Methods Phys. Res., Sect. B*, **288**, pp. 53–59.

Chapter 6

Simulation of a Lithium-Ion Battery

Takumi Yanagawa,[a] Hitoshi Sakagami,[b] and Kunioki Mima[c]
[a]*Department of Physics, Nagoya University, Furo-cho, Chikusa-ku, Nagoya, Aichi 464-8602, Japan*
[b]*National Institute for Fusion Science, National Institutes of Natural Sciences, 322-6 Oroshi-cho, Toki, Gifu 509-5292, Japan*
[c]*The Graduate School for the Creation of New Photon Industries, 1955-1 Kurematsu-cho, Nishi Ward, Hamamatsu, Shizuoka Prefecture, 431-1202, Japan*
sakagami.hitoshi@nifs.ac.jp

6.1 Introduction

An important task of lithium-ion battery simulations is to predict the lithium-ion battery characteristics by the experimental data obtained from various diagnostics like charge-discharge curve, ion beam analysis, X-ray diffraction, and neutron diffraction to the required precision with reasonable confidence. This task is typically performed by carrying out the simulations corresponding to the experiments and validating the simulation models of physicochemical processes that are observed in the diagnostic experiments. However, in a lithium-ion battery, most variables in the system such as local lithium-ion flux, negative ion flux in the electrolyte, lithium-ion

Lithium-Ion Batteries: Overview, Simulation, and Diagnostics
Edited by Yoshiaki Kato, Zenpachi Ogumi, and José Manuel Perlado Martin

ISBN 978-981-4800-40-2 (Hardback), 978-0-429-25934-0 (eBook)
www.panstanford.com

diffusivity, electrochemical potential, and local electrical resistivity, are not directly measurable during charge-discharge cycles in the experiments. Since the physical and chemical coefficients controlling the above quantities are necessary for reproducing the experimental data by the simulations, it is necessary to fully verify the accuracy of all of the physicochemical coefficients assumed in the simulations. The model coefficients that cannot be obtained directly have to be determined by comparing the experimental data with the model predictions. The battery physicochemical coefficients and operating conditions will be determined by a trial and error in combining the experiments and the simulations. By the above simulation modeling, the experimental data analysis and the design optimization can be made more efficiently. There are various simulations models for describing the microscopic phenomena and the macroscopic phenomena. In this chapter, a macroscopic simulation model and some simulations examples are described.

A mathematical simulation model for lithium-ion battery was firstly developed by J. Newman [1] and has been improved and extended by many researchers, for example, by A. Latz [2, 3]. The basic approach is to solve the transport of lithium ions in the electrolyte and inside active particles in electrodes. Figure 6.1 shows the schematic illustration of lithium-ion and electron transport in a lithium-ion rechargeable battery: from left to right, the components are a current collector, a positive electrode, a separator, a negative electrode, and a current collector. Both electrodes have active particles and conductive additives that are glued by a binder dampened with an electrolyte. The separator consists of an electrolyte and allows only ions (lithium ions and negative ions, not electrons) to pass through. During charging, lithium ions are precipitated on the surface of active primary particles in the positive electrode, which are transported in the electrolyte of the positive electrode, pass through the separator, and then intercalate in the active material of the negative electrode. When lithium ions are de-intercalated, electrons remain in active primary particles of the positive electrode that pass through active particles and conductive additives, and are then carried on the outside circuit from the current collector of the positive electrode. During discharging, on the other hand, the opposite directional transport of lithium ions and electrons occurs. The battery charge/discharge is carried out by transporting lithium ions between the positive electrode and the negative electrode.

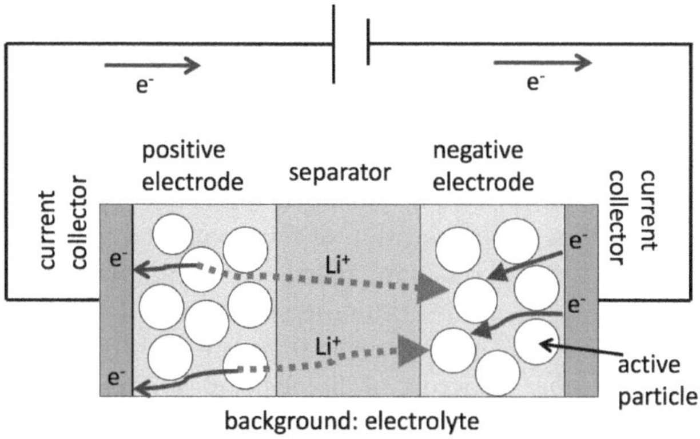

Figure 6.1 Lithium-ion and electron transport during charging. Empty circles in the electrodes represent secondary active particles that consist of nanometer-scale primary particles.

The electrochemical processes of a lithium-ion battery take place in various space–time scales. For the different space–time scale simulations, various mathematical models have been developed. The microscopic process of the lithium-ion battery occurs on the surface of the active particles where reactions take place in the electric double layer. The electrostatic structure and the Li ions' dynamics are simulated by Gaussian code describing quantum chemistry, the ab initio molecular dynamics, kinetic Monte Carlo, etc. The space–time scales of those models are shown in Fig. 6.2 [4].

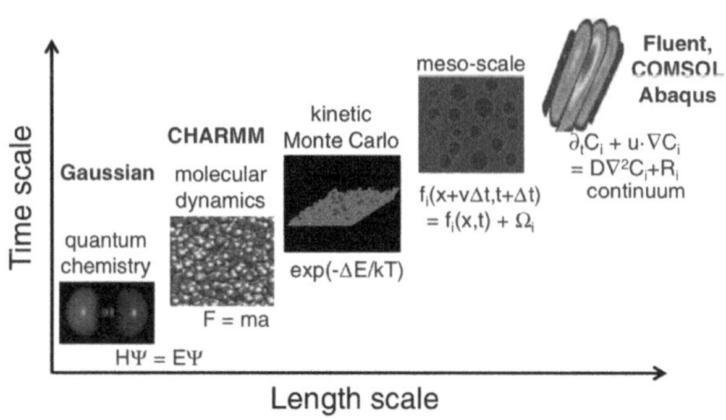

Figure 6.2 Approximate ranking of methods appropriate for the simulation of different times and scales. The details of the capability of the methods in this figure are described in the review paper by V. Ramadesigan, et al. [4].

The mesoscopic simulation model describes the behaviors of Li ions and the other ions as continuous media. The mesoscopic simulation model of the battery system during charge, discharge, and relaxation was developed by Newman et al. [1, 5, 6] and recently improved by Latz and Zausch et al. [2]. A theoretical model for the thermal effects on the lithium-ion battery has been given by Latz et al. [3] to describe the generation of hot spots and the localization of the electrochemical reaction, which are useful for simulating the durability and safety of the lithium-ion battery.

A typical 3D model simulation by Danner et al. [7] is shown in Fig. 6.3 for a given pore structure of the electrodes. Here, the potential, SOC, and Li distribution in the positive and negative electrodes are indicated by the gray scale when the battery is charged with a 1C rate. The 3D simulation result of Fig. 6.3 describes the Li drift and diffusion in the active particles, the intercalation/de-intercalation on the boundary between electrolyte and active materials, and diffusion inside active particles. In each area, the each element is an assumed continuum like a fluid. More details are described in the Ref. [7].

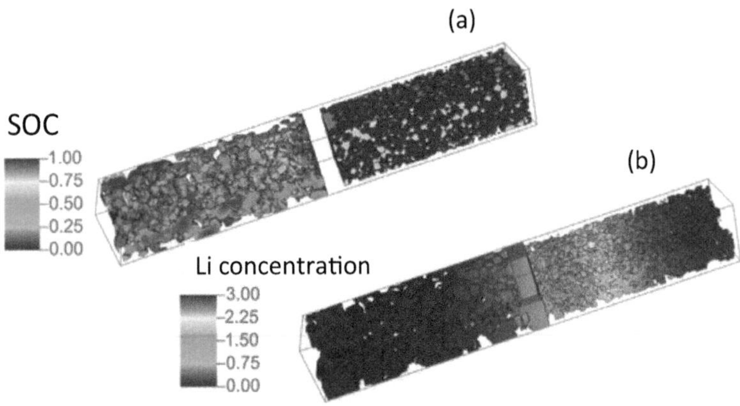

Figure 6.3 Graphs show (a) SOC and (b) Li concentration in a thick battery cell. Reprinted from Ref. [7], Copyright (2016) with permission from Elsevier.

The macroscopic models are convenient for analyzing the diagnostic results like ion beam analysis, since it is not necessary describe complicated pore structures and associating small-scale variations of the elemental concentration like Li concentration. In the simplified model, the electrodes are treated by two superimposed continua without regard for the actual geometric detail of the pore structure [8]. The separator consists of organic liquid acting as the solvent of a Li salt. The negative electrode is a lithium foil and the positive electrode is a porous electrode consisting of solid particles of the active material, a conducting filler, and an electrolyte. In the present model, it is assumed that the Li diffusion inside active particles is much faster than the charge/discharge characteristic time, and the condition for this assumption was given by Doyle and Newman [8] as follows:

$$S_e = \frac{{R_S}^2 I}{D_S F(1 - \varepsilon) c_T W_+} \ll 1, \tag{6.1}$$

where R_S is the radius of active particle in the positive electrode (m), I is the real current density (A/m^2), D_S is the diffusion coefficient of Li in the active particle (m^2/s), F is Faraday's constant, ε is porosity of the electrode, c_T is the maximum concentration of Li in an active particle (mol/m^3), and W_+ is the positive electrode thickness (m).

In this chapter, we describe the 1D continuum model in the following sections and the simulation results are compared with the experimental results taken by ion beam analysis.

6.2 Basic Equations

Lithium ions transport in electrolytes, while electrons transport in solid materials that are composed of secondary active particles and conductive additives. In this section, we introduce transport equations of lithium ions and electrons.

6.2.1 Transport of Lithium Ions in Electrolytes

The detail about transport equations of lithium ions in an electrolyte is described in articles written by Prof. Latz. The approach is as follows: From the thermodynamic equations for the electrochemical reactions, the entropy production is related to the thermal fluctuations that are electrothermodynamic forces and fluxes. Then, the entropy production is related to the decay of thermal fluctuations that are subjected to the fluctuation dissipation theorem (the Kubo formula). Then, the electrothermodynamic forces and fluxes must be linked by the Onsager reciprocal relation. Using the Onsager reciprocal relation and phenomenologically derived thermodynamic fluxes, the flux terms of transport equations are constituted. We will explain the process later on the basis of the articles by Prof. Latz.

As the first assumption, there is no convection in the electrolyte of the battery. Define ρ and $\mathbf{U}$ as mass density and flow velocity of the electrolyte, and they satisfy the continuous equation

$$\frac{\partial \rho}{\partial t} + \nabla \cdot \rho \mathbf{U} = 0. \tag{6.2}$$

In the case of no convection, the second term is

$$\nabla \cdot \rho \mathbf{U} = 0. \tag{6.3}$$

Therefore, the time derivative of mass density is

$$\frac{\partial \rho}{\partial t} = 0, \tag{6.4}$$

which means that the entire mass density does not vary with time. We consider the time variation of the mass densities of elements contained in the electrolyte, which are densities of the positive ion, negative ion, and neutral solvent. In the electrochemical field, molar mass is conventionally used instead of mass density. Following this convention, we define M_0, M_+, and M_- as the molar masses of the solvent, positive ion, and negative ion. The mass conservation Eq. 6.4) yields

$$M_0 dc_0 + M_+ dc_+ + M_- dc_- = 0, \tag{6.5}$$

where c_i is the molar concentration of the i-th element.

The second assumption is the charge neutrality condition between the positive ion and the negative ion:

$$z_+ c_+ + z_- c_- = 0, \tag{6.6}$$

where z_+ and z_- are the charge number of the positive ion and the negative ion, respectively. The charge separation like the electrical double layer occurs near the surface of active particle, but in the macroscale simulation, the charge separation on surfaces of the active material can be neglected [3]. Therefore, the divergence of the current should be zero:

$$\nabla \cdot \mathbf{j} = 0, \tag{6.7}$$

where $\mathbf{j}$ is the current density.

Note that the carriers of current j are the positive and the negative ions in the electrolyte. The electric conductivity is finite only in the active materials and the conductive additives.

We next give thermodynamic relations under the existence of electromagnetic fields to define entropy of the system.

$$de = Tds + \mu_+ dc_+ + \mu_- dc_- + \mu_0 dc_0 + \mathbf{E}\cdot d\mathbf{D} + \mathbf{H}\cdot d\mathbf{B}, \qquad (6.8)$$

where e, s, and μ_i are energy density, entropy density, and chemical potential of the kind of the i-th element, respectively, and T, $\mathbf{E}$, $\mathbf{D}$, $\mathbf{B}$, and $\mathbf{H}$ are temperature, electric field, electric flux density, magnetic flux density, and magnetic field, respectively. This equation represents variations of the internal energy in a small region. Note that pressure work PdV does not appear in this equation because the change of volume dV is zero due to the incompressibility. Equation 6.5 gives $dc_0 = -\dfrac{M_+dc_+ + M_-dc_-}{M_0}$, and using this relation and $c_- = -\dfrac{z_+}{z_-}c_+$ of Eq. 6.6, we can rewrite the terms related to chemical potential in Eq. 6.8 as

$$\mu_+ dc_+ + \mu_- dc_- + \mu_0 dc_0 =$$

$$dc_+\left\{\mu_+ - \frac{M_+}{M_0}\mu_0 - \frac{z_+}{z_-}\left(\mu_- - \frac{M_-}{M_0}\mu_0\right)\right\}. \qquad (6.9)$$

Define $c \equiv c_+$ and μ as effective chemical potential:

$$\mu \equiv \mu_+ - \frac{M_+}{M_0}\mu_0 - \frac{z_+}{z_-}\left(\mu_- - \frac{M_-}{M_0}\mu_0\right), \qquad (6.10)$$

Equation 6.8 is simplified as follows:

$$de = Tds + \mu dc + \mathbf{E}\cdot d\mathbf{D} + \mathbf{H}\cdot d\mathbf{B}. \qquad (6.11)$$

From the definition of entropy (e.g., p. 21 in Ref. [9]), the time evolution of entropy is represented as follows:

$$\frac{\partial s}{\partial t} = -\nabla\cdot\left(\frac{\mathbf{q}}{T}\right) + \frac{R}{T}, \qquad (6.12)$$

where $\mathbf{q}$ and R/T are heat flux and entropy production, respectively. The time derivative of Eq. 6.11 is given as follows:

$$\frac{\partial e}{\partial t} = T\frac{\partial s}{\partial t} + \mu\frac{\partial c}{\partial t} + \mathbf{E}\cdot\frac{\partial \mathbf{D}}{\partial t} + \mathbf{H}\cdot\frac{\partial \mathbf{B}}{\partial t}. \qquad (6.13)$$

Using Maxwell's equations, the last two terms of the right-hand side are represented as follows:

$$\mathbf{E}\cdot\frac{\partial \mathbf{D}}{\partial t}+\mathbf{H}\cdot\frac{\partial \mathbf{B}}{\partial t}=-\nabla\cdot(\mathbf{E}\times\mathbf{H})-\mathbf{j}\cdot\mathbf{E}. \tag{6.14}$$

Furthermore, the continuous equation about the positive ion is given as follows:

$$\frac{\partial c_+}{\partial t}=-\nabla\cdot\mathbf{N}_+. \tag{6.15}$$

with defining $\mathbf{N}_+$ as flux (molar concentration pass through unit area per unit time). Substituting Eqs. 6.12, 6.14, and 6.15 into Eq. 6.13, we obtain the following equation:

$$\frac{\partial e}{\partial t}=-\nabla\cdot(\mathbf{q}+\mathbf{N}_+\mu+\mathbf{E}\times\mathbf{H})+R+\mathbf{q}\cdot\frac{\nabla T}{T}-\mathbf{j}\cdot\mathbf{E}+\mathbf{N}_+\cdot\nabla\mu. \tag{6.16}$$

The conservation of total energy requires

$$\frac{\partial e}{\partial t}+\nabla\cdot\mathbf{I_E}=0, \tag{6.17}$$

where $\mathbf{I_E}$ is the energy flux. Comparing Eq. 6.16 with Eq. 6.17, energy flux $\mathbf{I_E}$ is obtained as follows:

$$\mathbf{I_E}=\mathbf{q}+\mathbf{N}_+\mu+\mathbf{E}\times\mathbf{H}. \tag{6.18}$$

Entropy production R/T must satisfy the following relation to satisfy the energy conservation law of Eq. 6.17:

$$R=-\mathbf{q}\cdot\frac{\nabla T}{T}-\mathbf{N}_+\cdot\nabla\mu+\mathbf{j}\cdot\mathbf{E}. \tag{6.19}$$

Here we rewrite the chemical potential with $\nabla\mu=\frac{\partial\mu}{\partial c}\nabla c+\frac{\partial\mu}{\partial T}\nabla T$ using c and T due to the difficulty of measurement of the chemical potential. Equation 6.19 can be rewritten as

$$R=-\mathbf{Q}\cdot\frac{\nabla T}{T}-\frac{\partial\mu}{\partial c}\mathbf{N}_+\cdot\nabla c+\mathbf{j}\cdot\mathbf{E}, \tag{6.20}$$

where $\mathbf{Q}=\mathbf{q}+\mathbf{N}_+T(\partial\mu/\partial T)$. The mathematical form of entropy production is now obtained. As shown next, the gradient term of the right-hand side corresponds to thermodynamic force (∇T and so on) and the coefficient corresponds to thermodynamic flux ($-\mathbf{Q}/T$ and so on).

Thermoelectric transport coefficients relate the fluxes to the thermodynamic forces, which are heat conductivity, electrical conductivity, diffusivity, thermoelectric effects, etc. You can easily imagine that the heat conductivity relates the heat flux to the temperature gradient as Fourier's law.

When multiple thermodynamic forces exist, however, it is not so simple, and the coefficients, which are connecting the forces and the fluxes, are represented as a matrix. The Onsager theorem proves that the coefficient matrix must be a symmetric matrix (the Onsager reciprocal relation) for positive entropy production, and this relation can be written as follows:

$$\begin{bmatrix} F_1 \\ F_2 \\ F_3 \end{bmatrix} = \begin{bmatrix} L_{11} & L_{13} & L_{13} \\ L_{21} & L_{22} & L_{23} \\ L_{31} & L_{32} & L_{33} \end{bmatrix} \begin{bmatrix} f_1 \\ f_2 \\ f_3 \end{bmatrix}. \tag{6.21}$$

where F_i and f_i are thermodynamic flux and thermodynamic force, respectively, and L_{ij} are the transport coefficients that relate the thermodynamic force to the conjugate flux. The Onsager reciprocity means $L_{ij} = L_{ji}$. This is reasonable because of the following reasons.

The fluctuation forces f_i are assumed to be small deviations from the thermal equilibrium, and the probability of the fluctuations is related to the entropy that is maximum at equilibrium. Therefore,

$$W = A\exp\left(\frac{S}{k_B}\right) \text{and} \frac{S}{k_B} = -1/2\beta_{ij} f_i f_j, \tag{6.22}$$

where k_B is the Boltzmann constant, $\beta_{ij} f_i f_j$ is the Einstein sum and nonnegative. Namely,

$$\beta_{ij} = \beta_{ji} = \frac{-1}{k_B} \frac{\partial^2 S}{\partial f_i \partial f_j}. \tag{6.23}$$

The entropy is produced by the decay of the fluctuations that are subjected to $f_i = -\lambda_{ij} f_j$, where all eigenvalues of λ_{ij} are positive if the system is stable. Then, the entropy production rate is given by

$$\frac{\partial S}{k_B \partial t} = \beta_{ij} \lambda_{jk} f_i f_k. \tag{6.24}$$

Let's introduce conjugate physical quantities, $F_i = \dfrac{-\partial S}{k_B \partial f_i} = \beta_{ij} f_j$.

Then the f_i of Eq. 6.24 is rewritten as $f_i = \beta_{ij}^{-1} F_j$ and the entropy production rate Eq. 6.24 has the following form:

$$\frac{\partial S}{k_B \partial t} = \gamma_{ij} f_i F_j, \tag{6.25}$$

where $\gamma_{ij} = \beta_{ik}\lambda_{kl}\beta_{lj}^{-1}$

In the Li-ion battery electrolyte, the entropy production is given as follows, according to Eq. 6.20:

$$\frac{R}{T} = \left\{ -\mathbf{Q} \cdot \frac{\nabla T}{T} - \frac{\partial \mu}{\partial c} \mathbf{N}_+ \cdot \nabla c + \mathbf{j} \cdot \mathbf{E} \right\} \times \frac{1}{T},$$

and comparing it with Eq. 6.25, we define the forces and the fluxes:

$$f_1 = \nabla C, f_2 = \mathbf{E}, f_3 = \nabla T,$$

$$F_1 = -\left(\frac{\partial \mu}{\partial c}\right)\mathbf{N}_+,\ F_2 = \mathbf{j},\ F_3 = -\frac{\mathbf{Q}}{T}. \tag{6.26}$$

The γ_{ij} can be also defined as the follows:

$$\gamma_{ii} = \frac{1}{k_B T}, \text{ for } i = 1, 2, 3,\ \gamma_{ij} = 0, \text{ when } i \neq j, \tag{6.27}$$

and $\gamma_{ii} L_{ij} = \beta_{il}\lambda_{lm}\beta_{mj}^{-1}$, namely $L_{ij} = \frac{1}{\gamma_{ii}} \beta_{il}\lambda_{lm}\beta_{mj}^{-1}$. Since β_{ij} is symmetric, the L_{ij} is symmetric if λ_{ij} is symmetric. The symmetry of λ_{ij} is indicated by the fluctuation dissipation theorem.

We obtain the coefficient matrix connecting the thermodynamic fluxes and the thermodynamic forces in the electrolyte:

$$\begin{bmatrix} \left(-\frac{\partial \mu}{\partial c}\right)\mathbf{N}_+ \\ \mathbf{j} \\ -\frac{\mathbf{Q}}{T} \end{bmatrix} = -\begin{bmatrix} L_{11} & L_{12} & L_{13} \\ L_{21} & L_{22} & L_{23} \\ L_{31} & L_{32} & L_{33} \end{bmatrix} \begin{bmatrix} \nabla c \\ \mathbf{E} \\ \nabla T \end{bmatrix}. \tag{6.28}$$

Because of the symmetric matrix, $L_{12} = L_{21}$, $L_{13} = L_{31}$, and $L_{23} = L_{32}$. Therefore, all flux terms are formulated if the diagonal elements, L_{12}, L_{13}, and L_{23} are determined.

Phenomenologically, the flux of the positive ion is represented as follows:

$$\mathbf{N}_+ = -D\nabla c + \frac{t_+}{z_+ F}\mathbf{j} - \frac{Dck_T}{T}\nabla T, \tag{6.29}$$

where D, t_+, and k_T are diffusivity, transference number, and Soret coefficient, respectively [10]. The transference number means the ratio of current contributed by the positive ion to total ionic current, which is also represented as $t_+ = \dfrac{m_+}{m_+ + m_-}$ where m_+ and m_- are the mobility of positive and negative ions, respectively.

The first term in Eq. 6.29 is the flux contributed by concentration diffusion, the second term is electric current, and the third term is thermal diffusion by temperature gradient (the Soret effect). Phenomenologically, electric current is represented as follows:

$$\mathbf{j} = x_1 \nabla c + \kappa \mathbf{E} - \beta\kappa \nabla T, \tag{6.30}$$

where κ and β are conductivity and the Seebeck coefficient, respectively, and x_1 is an unknown variable that is determined by the Onsager reciprocal relation. The first, second, and third terms are electric current by concentration gradient, Ohm's law, and the Seebeck effect known as the thermoelectric effect, respectively. Finally, phenomenological heat flux ($\mathbf{q}$) is represented as follows:

$$\mathbf{q} = x_2 \nabla c + x_3 \mathbf{E} - \lambda \nabla T, \tag{6.31}$$

where λ is thermal conductivity, and x_2 and x_3 are determined by the Onsager reciprocal relation. The last term means Fourier's law.

The unknown variables x_1, x_2, and x_3 can be determined by comparing Eqs. 6.29, 6.30, and 6.31 with the Onsager reciprocal relation. First, multiply $-\left(\dfrac{\partial \mu}{\partial c}\right)$ to Eq.(6.29) and substitute Eq. 6.30 into $\mathbf{j}$; then we obtain the equation as follows:

$$\begin{aligned} -\left(\frac{\partial \mu}{\partial c}\right)\mathbf{N}_+ &= (D - x_1)\left(\frac{\partial \mu}{\partial c}\right)\nabla c - \kappa \frac{t_+}{z_+ F}\left(\frac{\partial \mu}{\partial c}\right)\mathbf{E} + \\ &\left(\frac{t_+ \beta\kappa}{z_+ F} + \frac{Dck_T}{T}\right)\left(\frac{\partial \mu}{\partial c}\right)\nabla T. \end{aligned} \tag{6.32}$$

Comparing it with Eq. 6.28, one can find $L_{12} = -\kappa \dfrac{t_+}{z_+ F}\left(\dfrac{\partial \mu}{\partial c}\right)$. $L_{21} = x_1$ is also obtained by comparing Eq. 6.30 with Eq. 6.28. As $L_{12} = L_{21}$ should be satisfied from the Onsager reciprocal relation, we can get $x_1 = -\kappa \dfrac{t_+}{z_+ F}\left(\dfrac{\partial \mu}{\partial c}\right)$. As for the heat flux, we can get unknown variables in the same manner. In this book, we assume temperature is constant for the sake of simplicity. As a consequence, the transport equations related to lithium ions in the electrolyte as follows:

$$\left(\frac{\partial c}{\partial t}\right)+\nabla\cdot\mathbf{N}_{+}=0 \tag{6.33}$$

$$\mathbf{N}_{+}=-D\nabla c+\frac{t_{+}}{z_{+}F}\mathbf{j} \tag{6.34}$$

$$\mathbf{j}=-\kappa\nabla\Phi-\kappa\frac{t_{+}}{z_{+}F}\left(\frac{\partial\mu}{\partial c}\right)\nabla c \tag{6.35}$$

$$\nabla\cdot\mathbf{j}=0. \tag{6.36}$$

Here we introduce electric potential as $\mathbf{E} = -\nabla\Phi$. If transference number t_+ is constant, the transport equations of lithium ions are constituted as simple diffusion. In a dilute solution, the transference number is approximately treated as a constant. As shown in Section 6.3, the relation $\nabla \cdot \mathbf{j} = 0$ is not satisfied in the electrode where active particles exist in electrolyte, and not only the concentration but also $\mathbf{j}$ and Φ should be important in such a case.

6.2.2 Electron Transport in Active Particles and ConductIve Additives

Electron conduction is finite only in active particles and conductive additives, and electron movements are treated as an electronic current. Namely, electron transport follows Ohm's law:

$$\mathbf{j}_e = -\kappa_S\nabla\Phi_S, \tag{6.37}$$

where κ_S and Φ_S are conductivity of the electron and electric potential in the solid materials.

6.2.3 Electrochemical Reaction on the Surface of an Active Particle

During charge, lithium ions in the positive electrode are precipitated from the surface on the active particle by chemical reaction and are drained into the electrolyte. During discharge, the movement direction of lithium ions is opposite. The chemical reaction occurs on the surface of the active particle. Therefore the concentration of lithium ions is different between the surface and the center of the active particle. For the sake of simplicity, we neglect the lithium-ion transport inside the active particle by assuming the concentration distribution as rapidly uniform. This assumption is met if the condition of Eq. 6.1 is satisfied. This condition should be checked after the simulation for validating the simulation. The chemical reaction with the active particle and electrolyte is represented by the Butler–Volmer theorem, in which lithium-ion flux (electric current by lithium ions) at the surface is given by following equation [11]:

$$j_s = j_0 \{\exp(\alpha_a F\eta/RT) - \exp(-\alpha_c F\eta/RT)\}, \tag{6.38}$$

where α_a and α_c, which satisfy $\alpha_a + \alpha_c = 1$, are the weight factor of the contribution of anodic and cathodic reactions, respectively.

η is called the overpotential, that is, the potential gap between the surface of the active particle and the electrolyte, and is given by

$$\eta = \Phi_s + \frac{\mu_s}{z_+ F} - \left(\Phi + \frac{\mu}{z_+ F}\right). \tag{6.39}$$

j_0 is given as follows:

$$j_0 = k c_+^{\alpha_a} c_s^{\alpha_c} \left(1 - \frac{c_s}{c_{s,\max}}\right), \tag{6.40}$$

where k is the reaction rate constant, c_s is the lithium-ion concentration in the active particle, and $c_{s,\max}$ is the maximum concentration of lithium ions allowed to be stored in the active particle. In the case of considering a more complex chemical reaction model than the Butler–Volmer model, you may modify Eq. 6.38 with the estimated flux given by the new complex model. The lithium-ion flux obtained by the above process is the amount of lithium ions intercalated into/de-intercalated from the active particle, namely

$$\frac{\partial}{\partial t} \int_V c_s dV = \int_S \frac{j_s}{z_+ F} dS, \tag{6.41}$$

and $z_+ F$ is the factor for converting the lithium-ion flux to the lithium-ion current.

6.3 Macroscopic Simulation Method: The Continuum Model

As lithium ions transport only in electrolytes (electrolyte layer) and electrons transport only in solid materials (solid layer), we can treat the transports of lithium ions and electrons separately.

The lithium-ion transport in the electrolyte layer and the electron transport in the solid layer can be independently simulated by the use of basic equations derived in Section 6.2. But the lithium-ion source term is determined by the chemical reaction on the surface of active particles, which is induced by the electrochemical potential gap between the particle surface and the electrolyte. In our double-layer model, this potential gap is corresponding to the potential gap between the electrolyte and the solid layers and two layers are coupled through the chemical reaction of lithium ions. The schematic diagram of the double layer model is shown in Fig. 6.4. The Li source term is determined by the potential gap between two layers using the Butler–Volmer equation.

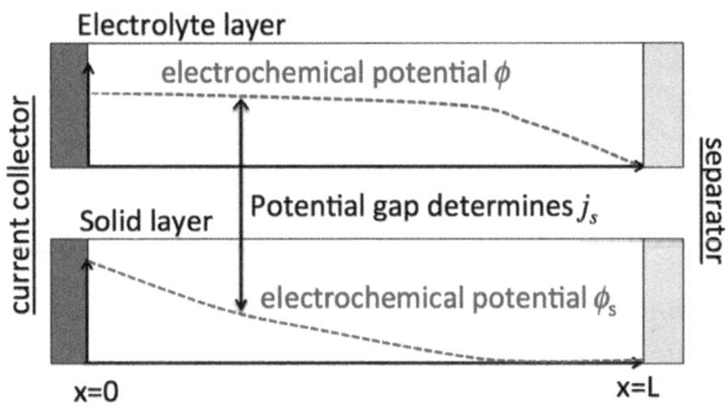

Figure 6.4 Schematic diagram of the double-layer model. Lithium ions transport only in the electrolyte layer, and electrons transport only in the solid layer. Two layers are coupled through the chemical reaction of lithium ions on the surface of active particles.

The most difficult part of the lithium-ion battery simulation is how to treat the electrode that is composed of active particles and electrolyte (and also conductive additives and binder). The electrode can be simulated with a microscopic and/or a macroscopic approach. The microscopic approach treats each active particle separately [12, 13]. This approach has the advantage of the large size of active particles, but it becomes quite difficult if the number of active particles increases and the size is small. On the other hand, the macroscopic approach uses coarse graining of active particles over space [14–16]. The macroscopic approach is effective when enough active particles exist in the electrode and their sizes are enough small compared with the size of the electrode. In this chapter we introduce the macroscopic approach.

6.3.1 Derivation of Continuous Model Equations

The macroscopic viewing allows us to assume that the active particle is small enough and distributes in the electrolyte uniformly. Figure 6.5 shows the illustration of the macroscopic view. The small region includes active particles and conductive additives, as shown in the microscopic diagram in the figure. In the simulation of macroscopic viewing, these small regions (we call specific volume) can be considered as a point. The governing equations of simulation are constructed by the use of physical values on these points. The physical value on each point is defined as the physical value averaged over the specific volume, namely using V_S as the specific volume, the physical value on each point is defined as follows:

$$\bar{\phi} = \frac{\int_{V_S} \phi dV}{V_S}. \tag{6.42}$$

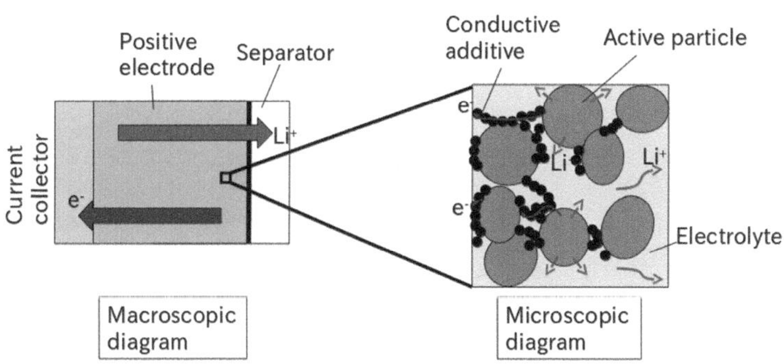

Figure 6.5 Illustration of a positive electrode with a macroscopic view and zoom-in illustration of a part of the electrode.

First, let's construct the continuous equation of lithium ions in the electrolyte. In the case of a pure electrolyte, there is no source term in the continuous equation, while in the case of an actual electrode, lithium ions are precipitated from active particles on each point; thus the lithium ions from active particles must be introduced as the source term in the continuous equation. Using Eq. 6.38, the amount of lithium ions de-intercaleted from or intercalated into active particles in the specific volume is represented as follows:

$$J_s = \int_{S_{\text{tot}}} j_s / z_+ F dS, \tag{6.43}$$

where S_{tot} is the total surface area of active particles in the specific volume. Defining the average value in the specific volume as

$$\bar{j}_s = j_s / S_{\text{tot}}, \tag{6.44}$$

the continuous equation of lithium ions in the electrolyte with the macroscopic view is given as follows:

$$\frac{\partial \bar{c}}{\partial t} + \nabla \cdot \bar{\mathbf{N}}_+ = (S_{\text{tot}} / V_s) \cdot \bar{j}_s / z_+ F, \tag{6.45}$$

where $\bar{c}$ and $\overline{\mathbf{N}_+}$ are $\int_{V_s} c dV / V_s$ and $\int_{V_s} N_+ dV / V_s$, respectively. S_{tot}/V_s is the fraction of the total surface area of all active particles occupied in the specific volume, which is the same physical value as the packing fraction of active particles per unit volume or porosity of porous particles. Since the source term Eq. 6.43 exists, charge neutrality Eq. 6.36 is not satisfied. Hence a source term should be added to Eq. 6.36 to satisfy the charge neutrality.

$$\nabla \cdot \bar{\mathbf{j}} = (S_{\text{tot}} / V_s) \bar{j}_s \tag{6.46}$$

When averaging $\boldsymbol{N}_+$, we used the relation $\overline{\nabla \cdot \mathbf{N}_+} = \nabla \cdot \overline{\mathbf{N}_+}$ because we assume $\boldsymbol{N}_+$ as follows:

$$\mathbf{N}_+ = \begin{cases} \mathbf{N}_+ & (\mathbf{r} \subset v_e) \\ 0 & (\mathbf{r} \subseteq v_a) \end{cases}, \tag{6.47}$$

where v_e and v_a are the region of the electrolyte and the active particle, respectively. In this case,

$$\begin{aligned} \int_{V_s} \nabla \cdot \mathbf{N}_+ / V_s dV &= \int_{v_e} \nabla \cdot \mathbf{N}_+ / V_s dV + \int_{v_a} \nabla \cdot \mathbf{N}_+ / V_s dV \\ &= \int_{v_e} \nabla \cdot \mathbf{N}_+ / V_s dV \end{aligned} \tag{6.48}$$

and, in $\mathbf{r} \subset v_e$, $\mathbf{N}_+$ is uniformly continuous,

$$\int_{v_e} \nabla \cdot \mathbf{N}_+ / V_s dV = \nabla \cdot \int_{v_e} \mathbf{N}_+ / V_s dV \tag{6.49}$$

is obtained. Thus $\overline{\nabla \cdot \mathbf{N}_+} = \nabla \cdot \overline{\mathbf{N}_+}$. On the other hand, without introducing the source term explicitly or assuming Eq. 6.47,

Eq. 6.45 can be derived. Averaging Eq. 6.34 in the specific volume,

$$\frac{\partial \bar{c}}{\partial t}+\overline{\nabla \cdot \mathbf{N}_+}=0. \tag{6.50}$$

Expand the second term of the left-hand side of Eq. 6.50 in the same manner as Eq. 6.48,

$$\frac{\partial \bar{c}}{\partial t}+\int_{v_e} \nabla \cdot \mathbf{N}_+ / V_s dV+\int_{v_a} \nabla \cdot \mathbf{N}_+ / V_s dV=0. \tag{6.51}$$

Applying Gauss's theorem to the third term of the left-hand side of Eq. 6.51, we can get

$$\frac{\partial \bar{c}}{\partial t}+\int_{v_e} \nabla \cdot \mathbf{N}_+ / V_s dV+\int_{s_a} \mathbf{N}_+ \cdot \mathbf{n} / V_s dS=0, \tag{6.52}$$

where $\mathbf{n}$ is the surface normal vector whose direction is inward because of the closed surface inside the boundary. As $\boldsymbol{N}_+ \cdot \mathbf{n}$ means $-j_s/z_+F$, therefore

$$\frac{\partial \bar{c}}{\partial t}+\nabla \cdot \overline{\mathbf{N}_+}=(S_{\text{tot}}/V_s)\bar{j}_s/z_+F \tag{6.53}$$

is derived. Here we used $s_a = S_{\text{tot}}$. This method is called the volume-averaging method [17, 18]. When a certain region includes a couple of phases (γphase and κ phase), the following relation holds in

$$\overline{\nabla \cdot \mathbf{F}}=\nabla \cdot \bar{\mathbf{F}}+\frac{1}{V}\int_{A_{\gamma\kappa}} \mathbf{F}_\gamma \cdot \mathbf{n}_{\gamma\kappa} dS, \tag{6.54}$$

where $A_{\gamma\kappa}$ represents the area of the $\gamma-\kappa$ interface contained within V. In Wang's article [14], the governing equations are derived by the use of the volume-averaging method. The volume-averaging method is a mathematical approach and the former case is a physical approach. In either case, same equations are derived, so we derive the governing equations from physical viewing.

Second, let's construct the continuous equation of lithium ions inside the active particles. As shown in Section 6.2.3 we assume the lithium-ion distribution becomes rapidly uniform in each active particle, so the continuous equation only consists of income and outcome of lithium ions:

$$\frac{\partial \bar{c}_s}{\partial t}=-(S_{\text{tot}}/V_s)\bar{j}_s/z_+F, \tag{6.55}$$

where $\overline{c}_s$ is the average value of the lithium-ion concentration in active particles inside the specific volume, $\int_{Vs} c_s dV/V_s$. As for the electronic current passing conductive additives, applying an averaging technique for current and potential at each point, namely $\int_{Vs} \mathbf{j}_e dV/V_s$ and $\int_{Vs} \Phi_s dV/V_s$, Eq. 6.37 is rewritten as follows:

$$\overline{\mathbf{j}_e} = -\kappa_s \nabla \overline{\Phi_s} \tag{6.56}$$

Furthermore, because the equivalent electrons with the lithium ions are produced in the specific volume, electronic current flows at each point in macroscopic view. Thus the source term is added to the electronic current in the same manner as Eq. 6.46.

$$\nabla \cdot \overline{\mathbf{j}_e} = -(S_{\text{tot}}/V_s)\overline{j_s}. \tag{6.57}$$

Note that the sign of electronic current is opposite to the electric current of lithium ions. Adding Eq. 6.46 to Eq. 6.57, one can find that the overall charge neutrality is satisfied.

Finaly, J_s at each point in the macroscopic view is obtained from the Butler–Volmer equation (Eq. 6.38) by the use of $\overline{\Phi}$ and $\overline{\Phi_s}$ defined at the point

$$\overline{j_s} = \overline{j_0}\,\{\exp(\alpha_a F\overline{\eta}/RT) - \exp(-\alpha_c F\overline{\eta}/RT)\}$$
$$\overline{\eta} = \overline{\Phi_s} + \frac{\overline{\mu_s}}{z_+F} - \left(\overline{\Phi} + \frac{\overline{\mu}}{z_+F}\right). \tag{6.58}$$

All governing equations of the electrode are derived as follows:

$$\frac{\partial c}{\partial t} + \nabla \cdot \mathbf{N}_+ = \varepsilon j_s/z_+F \tag{6.59}$$

$$\frac{\partial c_s}{\partial t} = -\varepsilon j_s/z_+F \tag{6.60}$$

$$\mathbf{N}_+ = -D\nabla c + \frac{t_+}{z_+F}\mathbf{j} \tag{6.61}$$

$$\mathbf{j} = -\kappa\nabla\Phi - \kappa\frac{t_+}{z_+F}\left(\frac{\partial\mu}{\partial c}\nabla c\right) \tag{6.62}$$

$$j_s = j_0\,\{\exp(\alpha_a F\eta/RT) - \exp(-\alpha_c F\eta/RT)\} \tag{6.63}$$

$$j_0 = kc_+^{\alpha_a} c_s^{\alpha_c}\left(1 - \frac{c_s}{c_{s,\max}}\right) \tag{6.64}$$

$$\eta = \Phi_s + \frac{\mu_s}{z_+F} - \left(\Phi + \frac{\mu}{z_+F}\right) \tag{6.65}$$

$$\mathbf{j_e} = -\kappa_s \nabla \Phi_s \tag{6.66}$$

$$\nabla \cdot \mathbf{j} = \varepsilon j_s \tag{6.67}$$

$$\nabla \cdot \mathbf{j_e} = -\varepsilon j_s. \tag{6.68}$$

Here $\varepsilon = S_{\text{tot}}/V_S$, and for the sake of simplicity, the bar symbol for average value is omitted. The unknown variables are c, c_S, and boundary conditions that are determined as current value in the case of constant current charging or voltage value in the case of constant voltage charging. Note that the chemical potential is not unknown variable, because it can be defined by the lithium-ion concentration and temperature.

Let's simplify the system equations from Eq. 6.59 to Eq. 6.68 before solving them numerically. Substituting Eq. 6.61 into Eq. 6.59 and then applying Eq. 6.67 gives

$$\frac{\partial c}{\partial t} = D\Delta c + \varepsilon j_s (1 - t_+) / z_+ F. \tag{6.69}$$

Taking divergence of Eqs. 6.62 and 6.66 and then applying Eqs. 6.67 and 6.68, the following relation is obtained:

$$\Delta \Psi = j_0 \varepsilon \left(\frac{1}{\kappa} + \frac{1}{\kappa_s} \right) \{\exp(\alpha_a F\eta/RT) - \exp(-\alpha_c F\eta/RT)\} + (t_+ / z_+ F)\Delta\mu, \tag{6.70}$$

where $\Psi = \Phi_S - \Phi$ and $\frac{\partial \mu}{\partial c}\nabla c = \nabla \mu$. Because $\eta = \eta(\Psi)$, Eq.(6.70) is a nonlinear differential equation. With given initial values and boundary conditions of c and c_S, and the boundary condition of Ψ, Eq. 6.70 can be solved and Ψ that satisfies the initial condition and the boundary condition is obtained. Once Ψ is obtained, η can be calculated and then j_S can be obtained. Thus we can solve Eq. 6.69.

The governing equations in separator region are equivalent to the equations that $j_S = 0$ from Eq. 6.34 to Eq. 6.36 or Eq. 6.69, because the separator consists of only the electrolyte.

$$\frac{\partial c}{\partial t} = D\Delta c. \tag{6.71}$$

Figure 6.6 shows the summary of the governing equations in each region of positive/negative electrodes and the separator. x_0,

x_p, x_n, and x_1 are positions of the boundary between the positive electrode and the current collector, between the positive electrode and the separator, between the separator and the negative electrode, and between the negative electrode and the current collector, respectively.

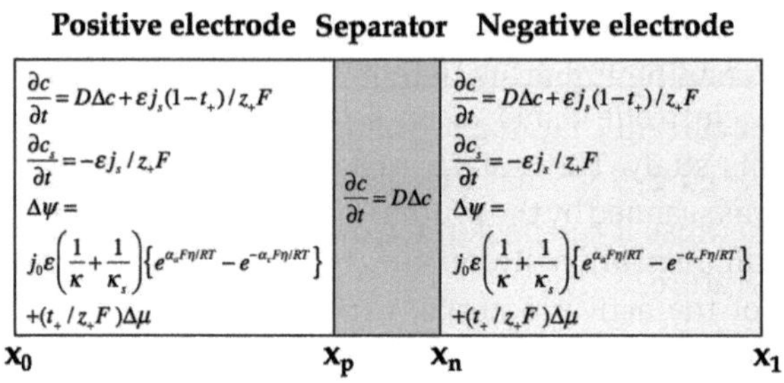

Figure 6.6 Governing equations of each region.

6.3.2 Discretization of 1D Governing Equations

Let's consider 1D governing equations for simplicity, namely $\frac{\partial}{\partial y} = \frac{\partial}{\partial z} = 0$. In this assumption, Eqs. 6.60, 6.69, 6.70, and 6.71 are as follows:

$$\frac{\partial c_s}{\partial t} = -\varepsilon j_s/z_+F(x_0 < x < x_p,\ x_n < x < x_1) \tag{6.72}$$

$$\frac{\partial c}{\partial t} = \begin{cases} D\frac{\partial^2 c}{\partial x^2} + \varepsilon j_s(1-t_+)/z_+F & (x_0 < x < x_p,\ x_n < x < x_1) \\ D\frac{\partial^2 c}{\partial x^2} & (x_p < x < x_n) \end{cases} \tag{6.73}$$

$$\frac{\partial^2 \Psi}{\partial x^2} = j_0\varepsilon\left(\frac{1}{\kappa}+\frac{1}{\kappa_s}\right)\left\{e^{\alpha_a F\eta/RT} - e^{-\alpha_c F\eta/RT}\right\} + (t_+/z_+F)\frac{\partial^2 \mu}{\partial x^2}(x_0 < x < x_p,\ x_n < x < x_1) \tag{6.74}$$

To solve differential equations numerically, we need to discretize continuous physical variables and transform differential equations into algebraic equations. The detail of the discretization is beyond the scope of this book and described in many textbooks about numerical calculations. There are several ways to discretize differential equations, but we adopt the finite differential method (FDM) here. In the FDM, the first time derivative of a certain function $g(t)$ is discretized with the forward differentiation as follows:

$$\left.\frac{dg(t)}{dt}\right|_{t=t_n} = \frac{g^{n+1} - g^n}{\Delta t} \tag{6.75}$$

$$g^n = g(t_n), \quad t_{n+1} = t_n + \Delta t.$$

The discretization of the second space derivative of a certain function $f(x)$ $(x_0 < x < x_1)$ is represented by the center differentiation as follows:

$$\left.\frac{d^2 f(x)}{dx^2}\right|_{x=x_i} = \frac{f_{i+1} - 2f_i + f_{i-1}}{\Delta x^2}$$

$$f_i = f(x_i), \quad x_i = x_0 + i\Delta x \quad (i = 0, ..., L) \tag{6.76}$$

$$\Delta x = (x_1 - x_0)/L,$$

where L is the number of partitions in the x direction. First, Eq. 6.72 is discretized as follows:

$$c_{s_i}{}^{n+1} = c_{s_i}{}^{n} - \Delta t \frac{\varepsilon j_{s_i}{}^{n}}{z_+ F}, \tag{6.77}$$

where time is represented by the upper suffix and space is represented by the lower suffix; namely, $c_{s_i}{}^n = c_s(x_i, t_n)$, $j_{s_i}{}^n = j_s(x_i, t_n)$. From Eq. 6.77, giving $c_{s_i}{}^n$ and $j_{s_i}{}^n$ at a time t_n, $c_{s_i}{}^{n+1}$ that means c_s after Δt in time can be algebraically calculated.

Second, Eq. 6.73 is also discretized in the same manner, and the discretization is given by

$$\frac{c_i{}^{n+1} - c_i{}^n}{\Delta t} =$$

$$\theta\left[D\frac{c_{i+1}^{n+1} - 2c_i^{n+1} + c_{i-1}^{n+1}}{\Delta x^2}\right] + (1-\theta)\left[D\frac{c_{i+1}^{n} - 2c_i^{n} + c_{i-1}^{n}}{\Delta x^2}\right] + S_i^n \tag{6.78}$$

$$S_i^n = \begin{cases} \varepsilon j_{s_i}^n (1-t_+)/z_+F & (x_0 < x < x_p,\ x_n < x < x_1) \\ 0 & (x_p < x < x_n) \end{cases},$$

where θ is an implicit parameter from 0 to 1. If $\theta = 0$, it is called the explicit method, while it is called the implicit method when $\theta \neq 0$. In this book, we use $\theta = 1/2$ and it is well known as the Crank–Nicolson method. With $\theta = 1/2$ and $\lambda_x = D\dfrac{Dt}{Dx^2}$, Eq. 6.78 is rewritten to

$$\begin{aligned} -\lambda_x c_{i-1}^{n+1} + 2(1+\lambda_x)c_i^{n+1} - \lambda_x c_{i+1}^{n+1} = \\ \lambda_x c_{i-1}^{n} + 2(1-\lambda_x)c_i^{n} + \lambda_x c_{i+1}^{n} + S_i^n \Delta t. \end{aligned} \quad (6.79)$$

In the case of $i = 0$ or $i = L$, Eq. 6.79 refers to the undefined point of $i = -1$ or $i = L + 1$. We define the physical value at these points by the use of the boundary condition. At the boundary, lithium ions can neither enter in nor leave from the outside. Therefor the lithium-ion flux is zero at the boundary, namely the gradient of the lithium-ion concentration must be zero, and the boundary condition of c is given by

$$\begin{aligned} \left.\frac{\partial c(x,t)}{\partial x}\right|_{x=x_0} &= \frac{c_0 - c_{-1}}{\Delta x} = 0 \\ \left.\frac{\partial c(x,t)}{\partial x}\right|_{x=x_L} &= \frac{c_{L+1} - c_L}{\Delta x} = 0. \end{aligned} \quad (6.80)$$

Equation 6.79 at $i = 0$ and $i = L$ is formed as

$$\begin{aligned} (2+\lambda_x)c_0^{n+1} - \lambda_x c_1^{n+1} &= (2-\lambda_x)c_0^n + \lambda_x c_1^n + S_0^n \Delta t \\ -\lambda_x c_{L-1}^{n+1} + (2+\lambda_x)c_L^{n+1} &= \lambda_x c_{L-1}^n + (2-\lambda_x)c_L^n + S_L^n \Delta t. \end{aligned} \quad (6.81)$$

Matrix representation of Eqs. 6.79 and 6.81 is as follows:

$$\begin{bmatrix} 2+\lambda_x & -\lambda_x & & & & & 0 \\ -\lambda_x & 2(1+\lambda_x) & -\lambda_x & & & & \\ & -\lambda_x & 2(1+\lambda_x) & -\lambda_x & & & \\ & & \ddots & \ddots & \ddots & & \\ & & & -\lambda_x & 2(1+\lambda_x) & -\lambda_x & \\ & & & & -\lambda_x & 2(1+\lambda_x) & -\lambda_x \\ 0 & & & & & -\lambda_x & 2+\lambda_x \end{bmatrix} \begin{bmatrix} c_0^{n+1} \\ c_1^{n+1} \\ c_2^{n+1} \\ \vdots \\ c_{L-2}^{n+1} \\ c_{L-1}^{n+1} \\ c_L^{n+1} \end{bmatrix} =$$

$$\begin{bmatrix} 2-\lambda_x & \lambda_x & & & & & 0 \\ \lambda_x & 2(1-\lambda_x) & \lambda_x & & & & \\ & \lambda_x & 2(1-\lambda_x) & \lambda_x & & & \\ & & \ddots & \ddots & \ddots & & \\ & & & \lambda_x & 2(1-\lambda_x) & \lambda_x & \\ & & & & \lambda_x & 2(1-\lambda_x) & \lambda_x \\ 0 & & & & & \lambda_x & 2-\lambda_x \end{bmatrix} \begin{bmatrix} c_0^{n} \\ c_1^{n} \\ c_2^{n} \\ \vdots \\ c_{L-2}^{n} \\ c_{L-1}^{n} \\ c_L^{n} \end{bmatrix} + \Delta t \begin{bmatrix} S_0^{n} \\ S_1^{n} \\ S_2^{n} \\ \vdots \\ S_{L-2}^{n} \\ S_{L-1}^{n} \\ S_L^{n} \end{bmatrix}. \quad (6.82)$$

Giving c_i^n and S_i^n, one can find that c_i^{n+1} can be calculated by solving the simultaneous equations in Eq. 6.82. Many numerical methods to solve the simultaneous equations $\mathbf{Ax} = \mathbf{b}$ are, for example, described in *Numerical Recipes* with program source codes. So you can choose any method that you want.

Finally, we consider $j_{s_i}{}^n$. To calculate $j_{s_i}{}^n$, we need to obtain Ψ_i^n by solving Eq. 6.74, which is discretized as follows:

$$\Psi_{i-1} - 2\Psi_i + \Psi_{i+1} = \Delta x^2 f(\Psi_i)$$

$$f(\Psi_i) = j_{0_i}\varepsilon\left(\frac{1}{\kappa}+\frac{1}{\kappa_s}\right)\left\{e^{\alpha_a F\eta(\Psi_i)/RT} - e^{-\alpha_c F\eta(\Psi_i)/RT}\right\} + (t_+/z_+F)\frac{\mu_{i-1}-2\mu_i+\mu_{i+1}}{\Delta x^2}, \tag{6.83}$$

where the upper suffix for time n is omitted. As variables except Ψ_i (i.e., j_{0_i} and μ_j) are given by the use of c_i^n and $c_{s_i}^n$ at t_n, the unknown variable is only Ψ_i. To solve Eq. 6.83, a boundary condition is required to be imposed. In the case of the constant current charging, the following boundary conditions are satisfied in the positive electrode $(x_0 < x < x_\text{p})$.

$$\begin{aligned}
j_\text{e}(x_0) &= -\kappa_\text{s}\left.\frac{d\Phi_\text{s}}{dx}\right|_{x=x_0} = j_\text{ext} \\
j(x_0) &= -\kappa\left.\frac{d\Phi}{dx}\right|_{x=x_0} - \frac{\kappa t_+}{z_+F}\left.\frac{d\mu}{dx}\right|_{x=x_0} = 0 \\
j_\text{e}(x_\text{p}) &= -\kappa_\text{s}\left.\frac{d\Phi_\text{s}}{dx}\right|_{x=x_\text{p}} = 0 \\
j(x_\text{p}) &= -\kappa\left.\frac{d\Phi}{dx}\right|_{x=x_\text{p}} - \frac{\kappa t_+}{z_+F}\left.\frac{d\mu}{dx}\right|_{x=x_\text{p}} = j_\text{ext},
\end{aligned} \tag{6.84}$$

where j_ext is the current density of an external current supply. Using the boundary condition of Eq. 6.84, the boundary conditions for Ψ are given by

$$\begin{aligned}
\left.\frac{d\Psi}{dx}\right|_{x=x_0} &= -\frac{j_\text{ext}}{\kappa_\text{s}} \\
\left.\frac{d\Psi}{dx}\right|_{x=x_\text{p}} &= \frac{j_\text{ext}}{\kappa} + \frac{t_+}{z_+F}\left.\frac{d\mu}{dx}\right|_{x=x_\text{p}}.
\end{aligned} \tag{6.85}$$

After discretizing Eq. 6.85, Eq. 6.83 at $i = 0$ and $i = i_p$ (i_p is the grid number at $x = x_p$) results in

$$\begin{aligned} -\Psi_0 + \Psi_1 &= \Delta x^2 f(\Psi_0) - \Delta x \frac{j_{ext}}{\kappa_s} \\ \Psi_{i_p-1} - \Psi_{i_p} &= \Delta x^2 f(\Psi_{i_p}) - \Delta x \left(\frac{j_{ext}}{\kappa} + \frac{t_+}{z_+ F} \frac{\mu_{i_p+1} - \mu_{i_p}}{\Delta x} \right). \end{aligned} \tag{6.86}$$

$\mu_{i_p} + 1$ in Eq. 6.86 can be defined by the use of $c_{i_p} + 1$ in the separator region because μ is a function of the lithium-ion concentration as $\mu(c)$. Let's represent Eqs. 6.83 and 6.86 as the matrix form with

$$J = \begin{bmatrix} -1 & 1 & & & & & 0 \\ 1 & -2 & 1 & & & & \\ & 1 & -2 & 1 & & & \\ & & \ddots & \ddots & \ddots & & \\ & & & 1 & -2 & 1 & \\ & & & & 1 & -2 & 1 \\ 0 & & & & & 1 & -2 \end{bmatrix}, \quad \Psi = \begin{bmatrix} \Psi_0 \\ \Psi_1 \\ \Psi_2 \\ \vdots \\ \Psi_{i_p-2} \\ \Psi_{i_p-1} \\ \Psi_{i_p} \end{bmatrix}$$

$$f(\Psi) = \begin{bmatrix} f(\Psi_0) \\ f(\Psi_1) \\ f(\Psi_2) \\ \vdots \\ f(\Psi_{i_p-2}) \\ f(\Psi_{i_p-1}) \\ f(\Psi_{i_p}) \end{bmatrix}, \quad \mathbf{B} = \begin{bmatrix} \dfrac{j_{ext}}{\kappa_s} \\ 0 \\ \vdots \\ \vdots \\ \vdots \\ 0 \\ \dfrac{j_{ext}}{\kappa} + \dfrac{t_+}{z_+ F} \dfrac{\mu_{i_p+1} - \mu_{i_p}}{\Delta x} \end{bmatrix}, \tag{6.87}$$

Equation 6.83 is simplified as follows:

$$J\Psi = \Delta x^2 f(\Psi) - \Delta x \mathbf{B}. \tag{6.88}$$

Equation 6.88 represents nonlinear simultaneous equations, so a certain linearization is required to solve it. One of the well-known methods to solve the nonlinear simultaneous equations is the Newton–Raphson method. In the Newton–Raphson method, with

$\delta\Psi$ being the difference between the predicted solution Ψ^* and the exact solution, Eq. 6.88 is expressed as follows:

$$J\,[\Psi^* + \delta\Psi] = \Delta x^2 f(\Psi^* + \delta\Psi) - \Delta x\mathbf{B}. \tag{6.89}$$

Applying the Taylor expansion to Eq. 6.89, we can get

$$f(\mathbf{\Psi}^* + \delta\mathbf{\Psi}) = \begin{bmatrix} f(\mathbf{\Psi}_0^*) + \delta\mathbf{\Psi}_0 \left.\dfrac{df}{d\mathbf{\Psi}}\right|_{\mathbf{\Psi}=\mathbf{\Psi}_0^*} \\ f(\mathbf{\Psi}_1^*) + \delta\mathbf{\Psi}_1 \left.\dfrac{df}{d\mathbf{\Psi}}\right|_{\mathbf{\Psi}=\mathbf{\Psi}_1^*} \\ \vdots \\ f(\mathbf{\Psi}_{i_p-1}^*) + \delta\mathbf{\Psi}_{i_p-1} \left.\dfrac{df}{d\mathbf{\Psi}}\right|_{\mathbf{\Psi}=\mathbf{\Psi}_{i_p-1}^*} \\ f(\mathbf{\Psi}_{i_p}^*) + \delta\mathbf{\Psi}_{i_p} \left.\dfrac{df}{d\mathbf{\Psi}}\right|_{\mathbf{\Psi}=\mathbf{\Psi}_{i_p}^*} \end{bmatrix} \tag{6.90}$$

$$= f(\mathbf{\Psi}^*) + f'(\mathbf{\Psi}^*)\,\delta\mathbf{\Psi},$$

and finally Eq. 6.88 results in

$$[J - \Delta x^2 f'(\mathbf{\Psi}^*)]\delta\mathbf{\Psi} = \Delta x^2 f(\mathbf{\Psi}^*) - J\mathbf{\Psi}^* - \Delta x\mathbf{B}, \tag{6.91}$$

where f' $f'(\mathbf{\Psi})$ Ψ is the diagonal matrix as follows:

$$f'(\mathbf{\Psi}) = \begin{bmatrix} \left.\dfrac{df}{d\mathbf{\Psi}}\right|_{\mathbf{\Psi}=\mathbf{\Psi}_0} & & & & 0 \\ & \left.\dfrac{df}{d\mathbf{\Psi}}\right|_{\mathbf{\Psi}=\mathbf{\Psi}_1} & & & \\ & & \ddots & & \\ & & & \left.\dfrac{df}{d\mathbf{\Psi}}\right|_{\mathbf{\Psi}=\mathbf{\Psi}i_p-1} & \\ 0 & & & & \left.\dfrac{df}{d\mathbf{\Psi}}\right|_{\mathbf{\Psi}=\mathbf{\Psi}i_p} \end{bmatrix}$$

Assuming an appropriate predicted solution Ψ^*, the simultaneous equations in Eq. 6.91 can be solved to obtain $\delta\Psi^*$. Using the obtained solution, update Ψ^* as $\Psi^* := \Psi^* + \delta\Psi^*$. Substituting a new solution Ψ^* into Eq.(6.91), $\delta\Psi^*$ is calculated again. This sequence is iterated until $\delta\Psi^*$ becomes infinitesimal. In general, when the norm of $\delta\Psi^*$ is small enough (<10^{-6}), Ψ^* is considered the exact solution. The way to find an appropriately predicted solution is heuristic, but it is good for convergence that the predicted solution is constructed by the solution one time step before; t_{n-1}. In the case of t_0, use Ψ satisfying $f(\Psi) = 0$ as the predicted solution.

As for the negative electrode ($x_n < x < x_1$), the same process is conducted. The difference from the positive electrode is the boundary condition, so we only show the boundary condition in the case of the negative electrode. The boundary condition corresponding to Eq. 6.84 is as follows:

$$\begin{aligned} j_e(x_n) &= -\kappa_s \left.\frac{d\Phi_s}{dx}\right|_{x=x_n} = 0 \\ j(x_n) &= -\kappa \left.\frac{d\Phi}{dx}\right|_{x=x_n} - \frac{\kappa t_+}{z_+ F}\left.\frac{d\mu}{dx}\right|_{x=x_n} = j_{ext} \\ j_e(x_1) &= -\kappa_s \left.\frac{d\Phi_s}{dx}\right|_{x=x_1} = j_{ext} \\ j(x_1) &= -\kappa \left.\frac{d\Phi}{dx}\right|_{x=x_1} - \frac{\kappa t_+}{z_+ F}\left.\frac{d\mu}{dx}\right|_{x=x_1} = 0. \end{aligned} \tag{6.92}$$

Thus Ψ of the positive electrode ($x_0 < x < x_p$) and the negative electrode ($x_n < x < x_1$) are obtained. Given Ψ, the overpotential η is calculated from Eq.(6.65). Substituting η into Eq.(6.63), j_s is obtained and the right-hand side of Eqs. 6.77 and 6.82 are determined. Finally, solving Eq. 6.77 and Eq. 6.82, $c_{s_i}{}^{n+1}$ and c_i^{n+1} are obtained. The flow of process is shown as follows:

1. Set the initial condition of c_i^n and $c_{s_i}{}^n$ at $n = 0$.
2. Solve the simultaneous Eqs. 6.83 and 6.86 in the positive and negative electrodes using c_i^n and $c_{s_i}{}^n$ and then obtain Ψ.
3. Substitute Ψ into Eq. 6.65 and then obtain η.
4. Substitute η into Eq. 6.63 and then obtain j_s.
5. Solve the system Eqs. 6.82 with j_s and then obtain c_i^{n+1}.
6. Solve Eq. 6.77 with j_s and then obtain $c_{s_i}{}^{n+1}$.
7. Redefine c_i^n and $c_{s_i}{}^n$ as $c_i^n := c_i^{n+1}$ and $c_{s_i}{}^n := c_{s_i}{}^{n+1}$ and go back to step 2 until t_{n+1} is the last time t_{end}.

6.4 Analysis of Experimental Results of Ion Beams by 1D Simulations

In this section, we show simulation results using the discretized equations of governing equations of a lithium-ion battery written in Section 6.3. Section 6.4.1 shows the simulation of a lithium-ion battery in charging by the use of typical parameters as an example. Section 6.4.2 shows the comparison of the lithium-ion profile obtained from particle-induced X-ray emission/particle-induced gamma-ray emission (PIXE/PIGE) experiments and simulations. Section 6.4.3 shows the simulation results about the relaxation process of the lithium-ion distribution when olivine-based material ($LiFePO_4$, called LFP subsequently) is used as the active particle of the positive electrode.

Table 6.1 Physical parameters for simulation

Parameter name	Positive electrode	Separator	Negative electrode
D [m^2/s] Diffusivity	25×10^{-10}	2.5×10^{-10}	2.5×10^{-10}
κ [S/m] Conductivity in electrolyte	2.5	2.5	2.5
κ_s [S/m] Conductivity in solid	10	–	100
t_+ Transference number	0.2	0.2	0.2
α_a/α_c Anodic/Cathodic weight	0.5	–	0.5
κ[$Am^{2.5}$–($Cmol^{0.5}$)] Reaction rate constant	1.0×10^{-6}	–	1.0×10^{-6}
ε [$1/m$] Specic surface area	1.0×10^8	–	1.0×10^8
$c_{s,max}$ [mol/m^3] Max. Lithium-ion concentration in active particle	5.1×10^4	–	3.1×10^4
μ^*[J/mol] Chemical potential in electrolyte at zero temperature	0.0	0.0	0.0
μ_s^* [J/mol] Chemical potential in active particle at zero temperature	-5.1×10^5	–	5.1×10^5

6.4.1 Charging Simulation

For carrying out simulations, it is necessary to provide all input parameters. They are categorized with those of physical property like diffusivity D, etc., and those of the simulation environment like time-step size Δt, etc. Parameters of the physical property are shown in Table 6.1 and parameters of the simulation environment are shown in Table 6.2. The physical property refers to Ref. [16]. The form of chemical potential in the electrolyte is approximated by that for a dilute solvent:

$$\mu = \mu^* + RT \log c, \tag{6.93}$$

where μ^* is the chemical potential in the electrolyte at zero temperature. On the other hand, we use the Nernst model for the form of chemical potential in active particles:

$$\mu_s = \mu_s^* + RT \log\left(\frac{x}{1-x}\right), \tag{6.94}$$

where $x = c_s/c_{s,\max}$ and μ_s^* is the chemical potential in the active particle at zero temperature.

Table 6.2 Simulation parameters

Parameter name	Positive electrode	Separator	Negative electrode
Thickness[m]	105×10^{-6} $(x_p - x_0)$	35×10^{-6} $(x_n - x_p)$	70×10^{-6} $(x_0 - x_n)$
$c(x, 0)$ [mol/m^3] Initial Lithium-ion concentration in electrolyte	1.0×10^3	1.0×10^3	1.0×10^3
$c_s(x, 0)$[mol/m^3] Initial Lithium-ion concentration in active particle	1.065×10^4	–	1.0×10^4
L Grid number	105	35	70
j_{ext} [A/m^2] Current density of constant current source		60	
z_+ Charge number		1	
R[J/K/mol] Gas constant		8.314	
T[K] Temperature		300	
Δx[m] Grid size		10^{-6}	
$D\frac{\Delta t}{\Delta x^2}$ Limitation of the time step size		1.0	

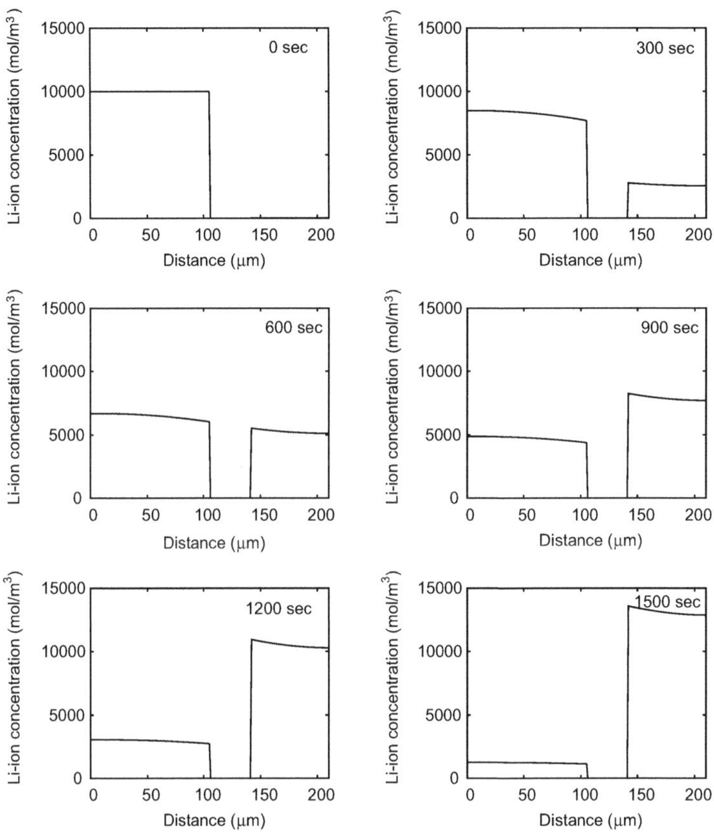

Figure 6.7 Time evolution of lithium-ion concentration in active particles (c_s).

In this table, the charge rate is assumed to be 2 C that is determined by the constant current supply, namely lithium ions are removed from the positive electrode in 30 min. Figure 6.7 shows the time evolution of the lithium-ion concentration in active particles (c_S). It is seen that 80% of lithium ions in the positive electrode ($0 < x < 105$) at $t = 0$ are transferred to the negative electrode ($140 < x < 210$). Furthermore the distribution of the lithium-ion concentration is non-uniform inside the electrodes. As it is known, non-uniformity depends on the charge rate and physicochemical properties of electrodes. Actually, in simulation results that will be shown in Section 6.4.3, the distribution of the lithium-ion concentration has a larger gradient when the charge rate is higher. Note that the lithium-ion concentration of the negative electrode can become larger than the initial concentration of the positive electrode because the negative electrode is thinner than the positive electrode.

Figure 6.8 shows the time evolution of the lithium-ion concentration in the electrolyte (c). The distribution of the lithium-ion concentration in the electrolyte varies early 20 s, but after 20 s the distribution is steady and hardly changed. This means that the relaxation time of lithium ions in the electrolyte contained in the separator is not short enough to compensate precipitation until 20 s.

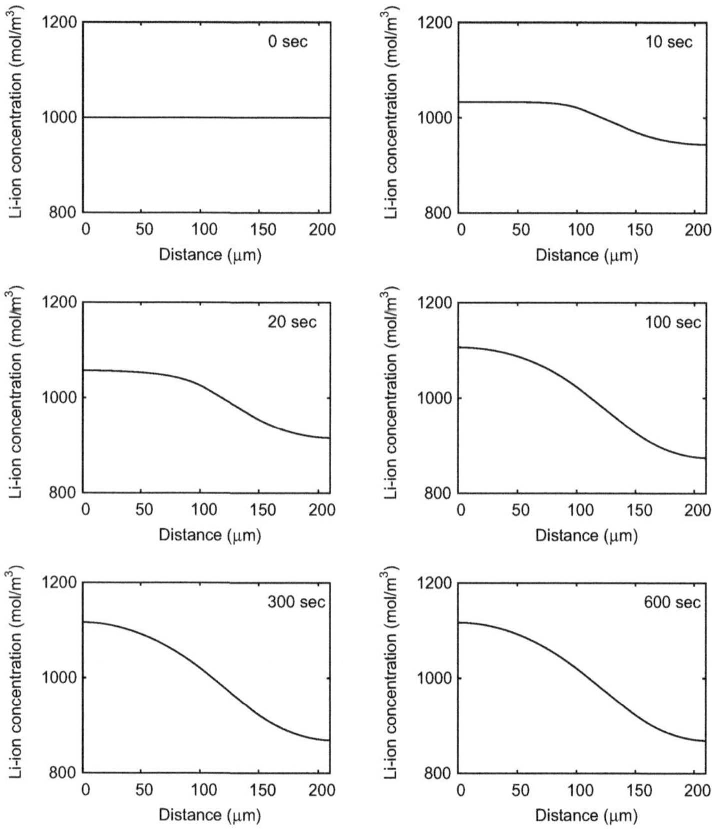

Figure 6.8 Time evolution of the lithium-ion concentration in the electrolyte (c).

Finally Fig. 6.9 and Fig. 6.10 show the time evolutions of the total lithium-ion amount of both electrodes and the voltage between both electrodes, respectively. The total amount of lithium ions of the positive electrode and the negative electrode is estimated by $\int_{x_0}^{x_p} c_s dx$ and $\int_{x_n}^{x_1} c_s dx$, respectively. From Fig. 6.9, it is found that the charge rate is 2 C and the total amount of lithium ions is conserved in the system. From Fig. 6.10, it is also found that the voltage of the battery is approximately 3V and the graph of the time evolution draws a typical charge curve. Initially, the voltage increases rapidly because of the non-uniform polarization of the separator, and then it is kept constant until approaching the full charge. The charge curve varies with charge conditions. In this simulation, we can obtain different curves by changing the current density of the constant current supply or the diffusivity of the electrolyte. Therefore, we can optimize the simulation parameters by fitting the voltage curves of simulations for various charge/discharge conditions with those of the experimental results.

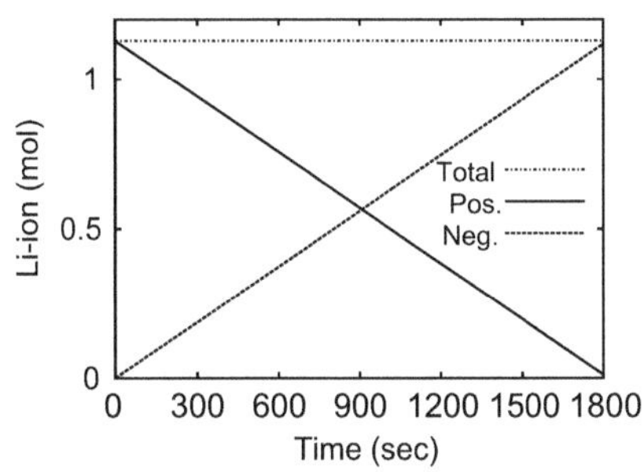

Figure 6.9 Time evolution of the total amount of lithium ion in active particles in both electrodes. Solid line: positive electrode; dashed line: negative electrode; dashed dotted line: sum of the positive and negative electrodes.

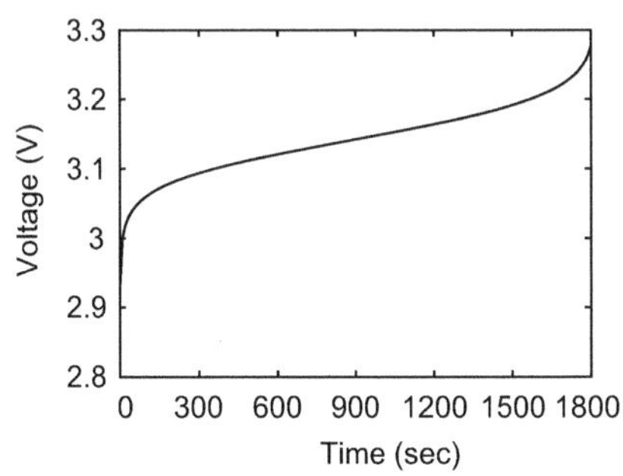

Figure 6.10 Time evolution of the voltage between both electrodes.

6.4.2 Electrode Thickness Dependence of the Lithium-Ion Distribution

Ion beam analysis of the lithium-ion battery by the use of TIARA shows that the distribution of the lithium-ion concentration has a smaller gradient in the thin electrode than that in the thick electrode. The simulations can reproduce the results of the ion beam analysis experiments.

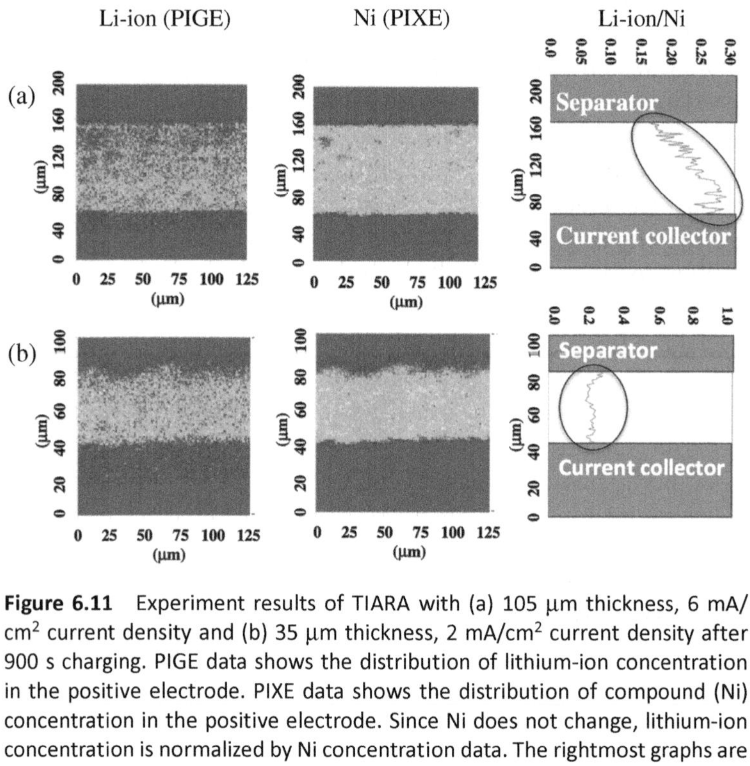

Figure 6.11 Experiment results of TIARA with (a) 105 μm thickness, 6 mA/cm^2 current density and (b) 35 μm thickness, 2 mA/cm^2 current density after 900 s charging. PIGE data shows the distribution of lithium-ion concentration in the positive electrode. PIXE data shows the distribution of compound (Ni) concentration in the positive electrode. Since Ni does not change, lithium-ion concentration is normalized by Ni concentration data. The rightmost graphs are 1D graphs where the distribution of the normalized lithium-ion concentration is averaged over horizontal direction of the 2D maps in the left-hand figures.

In the experiment, two types of electrodes are prepared where the thickness and charging current density of the positive electrode are 105 μm, 6 mA/cm^2 and 35 μm, 2 mA/cm^2, respectively. After a 900 s charge, the results of PIXE/PIGE are compared between these two samples. Figure 6.11 shows the result of TIARA, where the PIGE data is the distribution of the lithium-ion concentration in the positive electrodes and the PIXE data is the distributions of Ni in the positive electrodes. The right graph is a 1D graph averaged over the horizontal direction. From this graph, the gradient of concentration is larger in the case of 105 μm thickness than 35 μm, as shown by the circles in the rightmost pictures of Fig. 6.11.

Let's compare these experimental results with those of simulations. For simplicity, the same parameters with Section 6.4.1 are used, except thickness and current density of the constant current supply. Figure 6.12 shows the results of the lithium-ion concentration in active particles (c_s) charged after 900 s. Figure 6.12a is for 105 μm thickness of the positive electrode and 6 mA/cm^2 of the constant current supply, and Fig. 6.12b is for 35 μm thickness and 2 mA/cm^2 current density. From these results, the simulation also shows that the gradient of the lithium-ion concentration in the thin electrode is smaller than that in the thick electrode. These qualitative comparisons can be conducted easily and it is determined by the simulation model rather than simulation parameters whether the simulation result does or does not match with experimental results. The next section shows the simulation of the experiment that uses LFP as the positive electrode by modifying the chemical potential form.

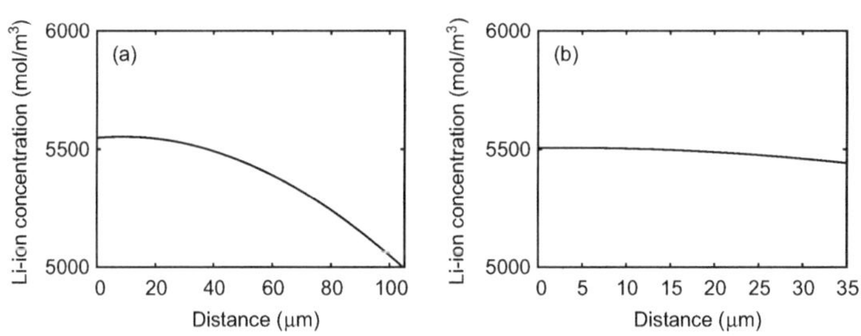

Figure 6.12 Lithium-ion concentration in active particles when thickness and current density are changed for (a) 105 μm thickness, 6 mA/cm^2 current density and (b) 35 μm thickness, 2 mA/cm^2 current density.

6.4.3 Simulation for an LFP Positive Electrode

Phosphoric acid iron lithium LFP does not comprise a rare metal and is economically attractive because of low-price active particles. In TIARA's experiments, relaxation of the distribution of the lithium-ion concentration for the positive electrode using LFP and $LiNiCoO_2$

(called LNO) is measured. In the case of LNO, the distribution of the lithium-ion concentration becomes uniform as time goes by; on the other hand, in the case of LFP, the distribution of the lithium-ion concentration hardly varies. A detailed explanation about the experiment is shown in Chapter 4. Figure 6.13 shows the lithium-ion concentration in active particles of the positive electrode for LNO and LFP after the end of charge. In the case of LFP, the distribution of lithium ions has a gradient even 1 h later. We will confirm these results by the simulation.

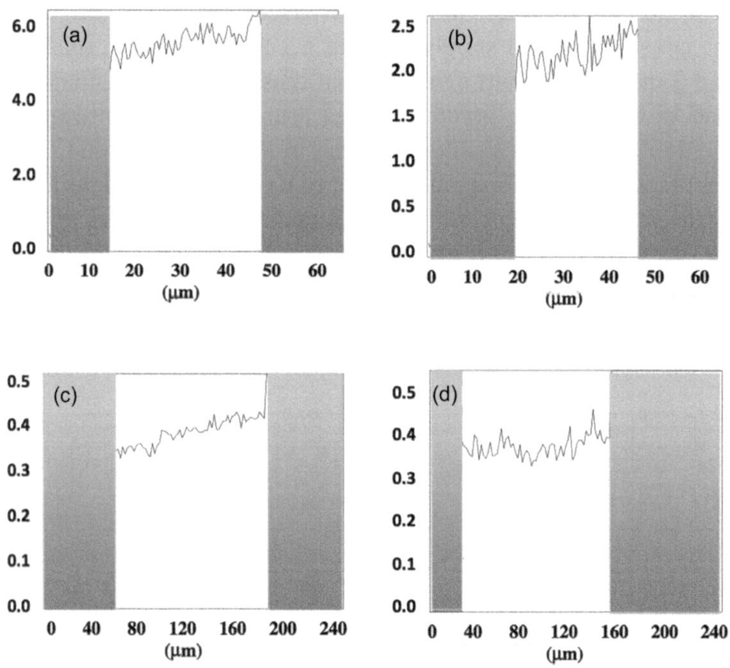

Figure 6.13 Relaxation process of distribution of the lithium-ion concentration after the end of charge: (a) 3 min. and (b) 1 h after finishing charge of LNO, and (c) 3 min. and (d) 1 h after finishing charge of LFP. The scales of the vertical axis are relative concentrations of lithium ions.

One of the most different features between Ni-based electrodes and Fe-based electrodes is the open-circuit voltage (OCV). In the case of LFP, the OCV is hardly subjected to the lithium-ion concentration. To reflect this tendency on our simulation, we modify the form of chemical potential of the positive electrode in our simulation model (Eq. 6.94).

$$\mu_s = \mu_s^* + \varepsilon_s RT \log\left(\frac{x}{1-x}\right), \tag{6.95}$$

where ε_s is a parameter that varies from 0 to 1. As ε_s is smaller, the dependency of chemical potential on the lithium-ion concentration is smaller. Simulations are conducted by the use of Eq. 6.95. Simulation parameters used are the same as those in Section 6.4.1. However, $D = 2.5 \times 10^{-11}$ for fitting the time scale of the experiment. The simulation is started by adding the gradient to c_s (distribution of the lithium-ion concentration in active particles of the positive electrode) and with current density of the constant current supply as zero.

Figure 6.14 shows the relaxation process of the distribution of the lithium-ion concentration in active particles of the positive electrode for ε_s = 1, ε_s = 0.1, and ε_s = 0.01. As ε_s becomes smaller, the relaxation speed becomes slower. This tendency agrees with the experiment result of LFP. Furthermore, this simulation result shows that ε_s in Eq. 6.95 determines the relaxation speed.

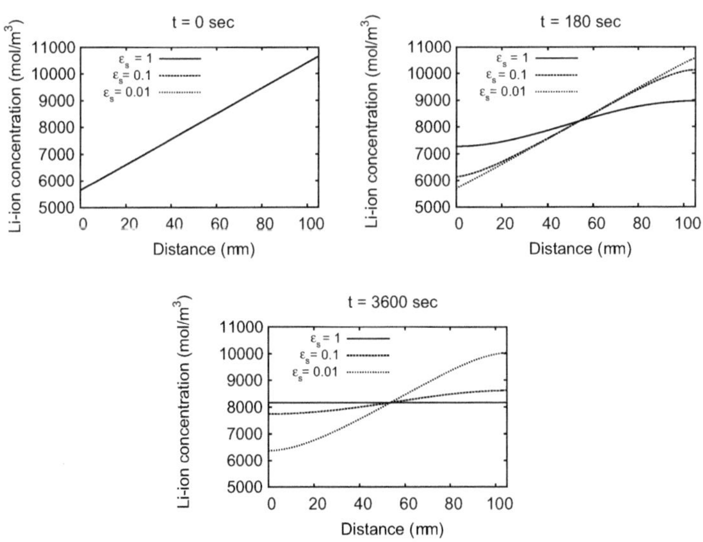

Figure 6.14 Relaxation process of the lithium-ion concentration in active particles of the positive electrode (c_s). Solid line: ε_s = 0; dashed line: ε_s = 0.1; dotted line: ε_s = 0.01.

Finally, we introduce a charge simulation using the form of chemical potential represented as Eq. 6.95. Figure 6.15 shows the time evolution of the lithium-ion concentration in active particles. The simulation parameters are the same as those in Section 6.4.1. It is found that the gradient of the concentration is larger compared to Fig. 6.7 and the gradient is the largest when $\varepsilon_s = 0.01$. This is caused by the difference of the relaxation speed, which means that the effective diffusion of lithium ions in active particles is small for $\varepsilon_s = 0.01$.

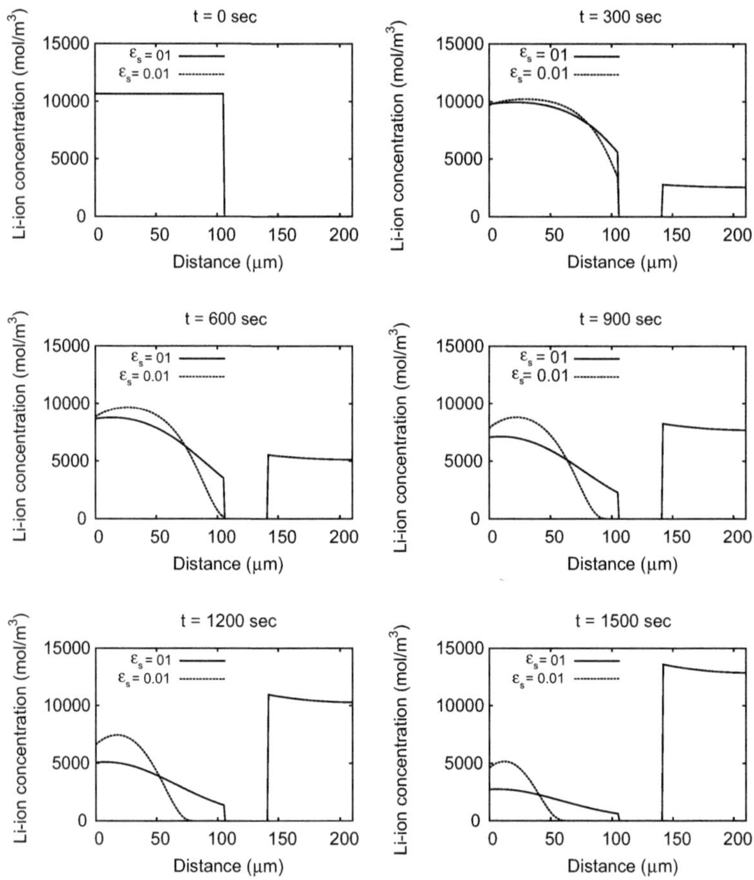

Figure 6.15 Time evolution of the lithium-ion concentration in active particles for LFP. Solid line: $\varepsilon_s = 0.1$; dashed line: $\varepsilon_s = 0.01$.

6.5 Summary

Macroscopic simulations of a lithium-ion battery were conducted by the use of the lithium-ion transport equations. The time evolution of the lithium-ion concentration in electrodes was simulated, and the distribution of the lithium-ion concentration was investigated by changing parameters. There are many parameters in simulations; thus you can choose appropriate parameters as matching with experimental results like charge/discharge properties, or you can determine unclear physical property in experiments by fitting simulation results with experimental results. From the point of view, in situ experiments that measure the distribution of the lithium-ion concentration in real time during charging will be powerful tools, and ion beam analysis is one of the candidate methods for in situ experiments of the lithium-ion battery.

In this chapter we introduced only 1D simulations, but the expansion to multidimensional simulation can be implemented by the same manner. In the case of multidimensions, the boundary condition often becomes a pure Neumann condition. In a pure Neumann condition, the solution cannot be obtained unless an appropriate numerical method for simultaneous equations is chosen.

References

1. Fuller, T. F., Doyle, M. and Newman, J. (1994a). Simulation and optimization of the dual lithium ion insertion cell, *J. Electrochem. Soc.*, **141**, pp. 1–10.
2. Latz, A., Zausch, J. and Iliev, O. (2010). Modeling of species and charge transport in Li–ion batteries based on non-equilibrium thermodynamics, *Berichte des Fraunhofer ITWM*, **190**, pp. 1–8.
3. Latz, A. and Zausch, J. (2011). Thermodynamic consistent transport theory of Li-ion batteries, *J. Power Sources*, **196**, pp. 3296–3302.
4. Ramadesugan, V., Northrop, P. W. C., De, S., Santhanagopalanb, S., Braatzc, R. D. and Subramaniana, V. R. (2012). Modeling and simulation of lithium-ion batteries from a systems engineering perspective, *J. Electrochem. Soc.*, **159**, pp. R31–R45.

5. Doyle, M., Fuller, T. F. and Newman, J. (1993). Modeling of galvanostatic charge and discharge of the lithium/polymer/insertion cell, *J. Electrochem. Soc.*, **140**, pp. 1526–1533.
6. Fuller, T. F., Doyle, M. and Newman, J. (1994b). Relaxation phenomena in lithium-ion-insertion cells, *J. Electrochem. Soc.*, **141**, pp. 982–990.
7. Danner, T., Signh, M., Latz, A., et al. (2016). Thick electrodes for Li-ion batteries: a model based analysis, *J. Power Sources*, **334**, pp. 191–201.
8. Doyle, M. and Newman, J. (1995). Modeling the performance of rechargeable lithium-based cells: design correlations for limiting cases, *J. Electrochem. Soc.*, **54**, pp. 46–51.
9. Kitahara, K. (1999). *Statistical Mechanics of Non-Equilibrium System* (Iwanami Shoten).
10. de Groot, S. and Mazur, P. (1962). *Non-Equilibirium Thermodynamics* (North-Holland).
11. Newman, J. and Thomas-Alyea, K. E. (2004). *Electrochemical Systems* (Wiley).
12. Popov, P., Vutov, Y., Margenov, S. and Iliev, O. (2010). Finite volume discretization of equations describing nonlinear diffusion in Li-ion batteries, *Berichte des Fraunhofer ITWM*, **191**, pp. 1–9.
13. Taralov, M., Taralova, V., Popov, P., Iliev, O., Latz, A. and Zausch, J. (2012). Report on finite element simulations of electrochemical processes in Li-ion batteries with thermic effects, *Berichte des Fraunhofer ITWM*, **221**, pp. 1–31.
14. Wang, C. Y. and Gu, W. B. (1998). Micro-macroscopic coupled modeling of batteries and fuel cells, *J. Electrochem. Soc.*, **145**, pp. 3407–3417.
15. Doyle, M., Fuller, T. F. and Newman, J. (1994). The importance of the lithium ion transference number in lithium/polymer cells, *Electrochem. Acta*, **39**, pp. 2073–2081.
16. Ning, G., White, R. E. and Popov, B. N. (2006). A generalized cycle life model of rechargeable Li-ion batteries, *Electrochem. Acta*, **51**, pp. 2012–2022.
17. Whitaker, S. (1999). *The Method of Volume Averaging* (Springer Science and Business Media).
18. Hjertager, B. H. (2007). *Multi-fluid CFD Analysis of Chemical Reactors* (Springer).

Chapter 7

Future Prospects of Intense Laser-Driven Ion Beams for Diagnostics of Lithium-Ion Batteries

Shunsuke Inoue, Masaki Hashida, and Shuji Sakabe
Advanced Research Center for Beam Science, Institute for Chemical Research, Kyoto University, Gokasho, Uji, Kyoto 611-0011, Japan
sinoue@lasear.kuicr.kyoto-u.ac.jp

7.1 Introduction

The recent development of ultrashort high-power laser systems has been encouraging the study of intense laser matter interactions to explore the potential for many attractive applications, for example, charged-particle acceleration [1–3], fast ignition for inertial confinement fusion [4, 5], ultrafast electron diffraction measurement [6, 7], time-resolved X-ray proving [8, 9], laser-driven nuclear physics [10], and laboratory astrophysics [11]. The acceleration of multi-mega–electron volt protons through the interaction of an ultraintense laser pulse with a thin foil is one of the

Lithium-Ion Batteries: Overview, Simulation, and Diagnostics
Edited by Yoshiaki Kato, Zenpachi Ogumi, and José Manuel Perlado Martin

ISBN 978-981-4800-40-2 (Hardback), 978-0-429-25934-0 (eBook)
www.panstanford.com

most interesting research issues because of the compactness and unique characteristics of proton beams. The mechanism of proton acceleration is called target normal sheath acceleration (TNSA) [12, 13]. First at the front surface of the foil, target electrons are accelerated to high energy by an intense laser pulse, pass through the target without collision in the target plasma, and are emitted into vacuum. Consequently, at the rear surface of the target, an ultrahigh electric field (sheath field) is generated. This electrostatic field is considered to be $\sim 10^{12}$ V/m in strength [13] and as short as an incident laser pulse in duration [14]. The sheath field ionizes atoms (for instance hydrogen) and accelerates the ions (protons) normal to the target surface in a small (typically about 1 to 10 μm) area [12, 13]. The proton beam generation by the TNSA has been intensively studied because the hydrogen atoms are abundantly included in any target foils and the lightest to be accelerated. In what follows, we describe "protons" for ions. The emittance of the energetic proton beam is mainly determined by the shape of the target, and ultralow emittance has been observed [15–17]. The emitted proton beam is additionally accelerated at the plasma expansion front [17, 18]. More than 10^{13} particles of energetic protons are observed in the energy of >10 MeV by a single laser pulse and the maximum observed energy is 58 MeV [2] at this moment. These remarkable features (ultralow emittance, high particle numbers per bunch, and short pulse duration) are offering many attractive applications, such as proton radiography [18, 19], high-energy-density matter [20], cancer therapy [21], and proton fast ignition [22].

As described above, there are many studies of laser-accelerated ion beams. So far there are few research studies for the usage of the laser-accelerated ion beams for ion beam analysis (IBA). It is thought that the reason there are few studies is the specifications of laser-accelerated ion beams (e.g., total integrated flux, stability, and divergence) being realized now. This chapter discusses the use of laser-accelerated ion beams in IBA. Through this chapter, we attempt to contribute toward realizing a society where the people can have reasonable and easy access to IBA applications.

7.2 Generation of Ion Beams by High-Intensity Lasers

In this section, the progress of the ultrafast intense laser science is introduced and the physics of interactions between ultrafast intense laser pulses and solid matters are briefly discussed to understand the mechanism of ion beam acceleration with intense laser pulses. Then one of the main and effective mechanisms for accelerating mega-electron-volt-energy ion beams, that is, the TNSA mechanism, is discussed.

7.2.1 Intense Ultrashort Lasers

Since the invention of laser in 1960, abundant laser applications have prevailed in many fields and laser technologies have seen remarkable development. Progress of ultrashort pulse laser technology with the invention of chirped pulse amplification (CPA) [15, 16] has been increasing peak power and intensity of laser pulses dramatically (Fig. 7.1). The laser pulse duration from a CPA laser system has been decreased to tens of femtoseconds. Focusing the laser pulses on an area a few micrometers in diameter heightens laser fields, which can accelerate electrons directly as fast as relativistic velocity. Irradiating matter with such an intense laser pulse instantaneously produces high-energy-density plasma in a microarea.

From laser plasmas, various radiations, such as electrons [23, 24], ions [12, 25], X-rays [26, 27], γ-rays [28], and terahertz waves [29, 30], are generated. These radiations have desirable features, such as short pulse duration, high intensity, point source, and perfect synchronization of different radiations from the laser plasma with each other. The radiations from laser plasmas have a high potential for attractive applications in many areas, for example, particle acceleration [1–3], fast ignition for inertial confinement fusion [4, 5], cancer therapy using ion beams [21], ultrafast electron diffraction measurement [6, 7], time-resolved X-ray proving [8, 9], laser-driven nuclear physics [10], and laboratory astrophysics [11].

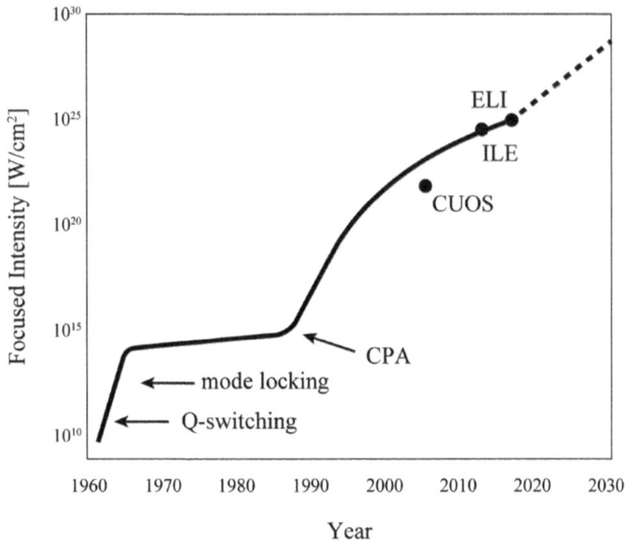

Figure 7.1 Time evolution of laser intensity. The focused intensity has jumped up with CPA innovation [40].

7.2.2 Generation and Transport of Fast Electrons on Interaction of an Intense Ultrashort Laser Pulse with Solid Matter

When solid matter is irradiated with an intense laser pulse, electrons in the target are accelerated to high energy (relativistic speed). The accelerated electrons, called "fast electrons," have an average energy 1 or 2 orders of magnitude higher than the electrons thermalized by collisional processes in plasma and induce laser plasma radiations (here, "radiation" is in the broad sense of the term, that is, both charged particles and electromagnetic waves, such as X-ray, ions, THz wave, and higher harmonics).

The generation mechanism and characteristics of the laser-accelerated proton beams are deeply related to the fast electron dynamics (the temporal and spatial characteristics of fast electrons) during or immediately after the laser interaction. To realize the applications mentioned above, it is crucial to understand the mechanisms of generation and transport of laser-accelerated fast electrons in plasmas.

Some theoretical interpretations have been proposed for the acceleration of fast electrons in the laser plasma interactions [31–34]. In Figs. 7.2, 7.3, and 7.4, the schematics of the typical theories are shown. Figure 7.2 shows "resonance absorption." When a p-polarized laser pulse is irradiated obliquely onto the target material the density of which has gradient along the z direction in Fig. 7.2 and the incident laser pulse reflects at the density below the critical density n_c, the electric field has a singularity at n_c (of the critical surface). The electric field is schematically shown in Fig. 7.2. This electric field resonantly oscillates electrons along the z direction at the reflection point and causes charge separation. This electron oscillation is damped in the plasma with collisional or collisionless processes; consequently, the target is heated by the laser pulse.

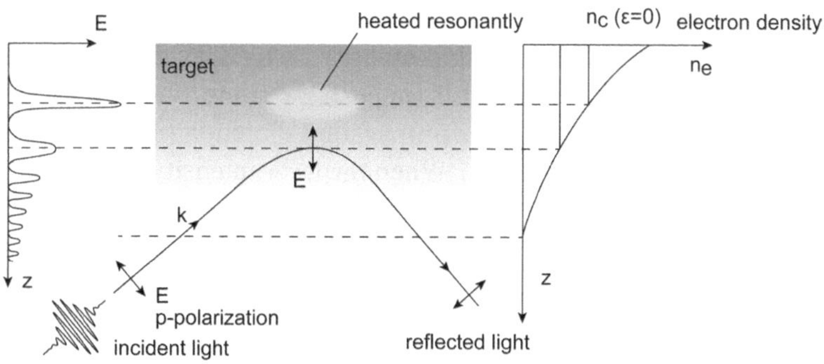

Figure 7.2 Schematic description of the resonance absorption model [17].

Figure 7.3 shows Brunel absorption (vacuum heating) [32]. This absorption also occurs on irradiation of a p-polarized laser pulse. In plasma, electrons are oscillated by the electric field of the laser pulse **E** with the maximum velocity of $\mathbf{v}_{osc} = e\mathbf{E}/m\omega$, where e is the elemental charge, m is the electron mass, and ω is the laser frequency. When the laser field oscillates electrons along the z direction around the critical surface, the electrons are pulled out into the vacuum region and pushed back to the plasma during half a cycle of the incident laser. If the density gradient is steep and during the next half cycle the laser field cannot exist in the region where the electron density is higher than the critical density, the electrons are emitted toward the target with a velocity of $\mathbf{v}_{osc}$; consequently, the energy of the laser pulse is transferred to the electrons.

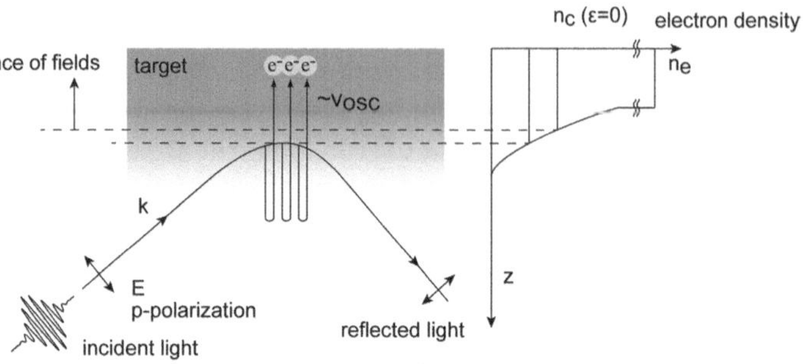

Figure 7.3 Schematic description of the Brunel absorption model (vacuum heating) [12].

Figure 7.4 shows the schematic of **j** × **B** heating [34]. When the laser intensity is not so high, the magnetic field **B** does not influence the motion of electrons and electrons are only quivered by the electric field. In a high-intensity laser field, however, the magnetic field of the laser pulse can push electrons oscillated by the electric field toward the laser propagation direction (due to the $e\mathbf{v} \times \mathbf{B}$ term of the Lorentz force). When the laser intensity is over 10^{18} W/cm^2, the velocity of electrons accelerated by the magnetic field becomes relativistic. When such a high-intensity laser irradiates the target with a steep density gradient on the surface, the electrons accelerated by the laser field are emitted into the target.

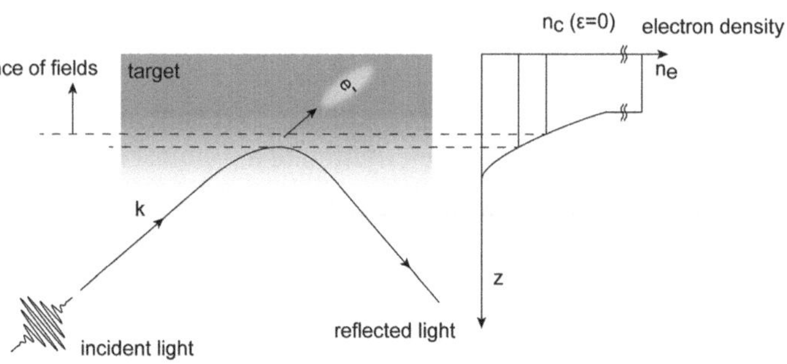

Figure 7.4 Schematic description of the **j** × **B** heating model [47].

The energy of the laser pulses is transferred to the fast electrons through the above-mentioned mechanisms. The actual process for accelerating fast electrons is rather complicated to understand with a simple model because the absorption processes that occur are complex and depend on various parameters. The total converted energy of the fast electrons is empirically scaled with the intensity of the incident laser pulses. At the intensity of 10^{18} W/cm^2, about 10% energy of the incident laser pulse is transferred to the fast electrons [35]. The energy spectrum of the fast electrons is also scaled with the intensity of the incident laser. It is known that they have Maxwellian form with a characteristic temperature T_h [36]. For intensity >10^{18} W/cm^2, which is the required intensity to accelerate the ions to mega–electron volt energy, the $\mathbf{j} \times \mathbf{B}$ heating is dominant and the temperature of the fast electrons T_h is written as T_h (keV) ≈ 511 $[(1 + 0.73I\lambda^2)^{1/2} - 1]$, where I is the intensity of the laser pulse in units of 10^{18} W/cm^2 and λ is the laser wavelength in micrometers.

7.2.3 Target Normal Sheath Acceleration

When an intense laser pulse is irradiated on a foil target and electrons are accelerated as described above, the fast electrons cross the foil target to the other side. When the fast electrons travel through the boundary of the foil and vacuum (rear surface of the foil target), a strong quasi-static electric field is created at the boundary due to the charge separation (like capacitance). This electric field is initially generated by the front of the fast electron pulse. The typical magnitude of this electric field reaches up to 10^{12} V/m. The strong electric field traps the following fast electrons, and they are deflected or returned back into the foil target. A part of these captured electrons leaks into the vacuum (this field is called "electron sheath"), and then the strong electric field is maintained while the fast electrons are supplied. This quasi-static electric field can easily ionize atoms around the rear surface of the foil and accelerates them toward normal direction of the foil surface. The accelerated particles are mainly protons, which arise from water vapor and hydrocarbon contamination. This mechanism is called the TNSA model (Fig. 7.5) [12].

The TNSA-accelerated proton beams have unique features in comparison with conventional particle accelerators. Some of the significant features are listed below.

- TNSA proton beams have ultralow transverse emittance, below 10^{-2} mm mrad [17].

- The bunch duration of the TNSA proton beams is a few picoseconds, which depends on the duration of the electric field. The bunch duration becomes longer due to their energy spread (described later).
- It is reported that the particle number in one bunch is more than 10^{13} for energy >10 MeV [2].
- The acceleration field (10^{12} V/m) is several orders of magnitude greater than conventional accelerators. The TNSA mechanism has potential for drastically reducing the accelerator size.

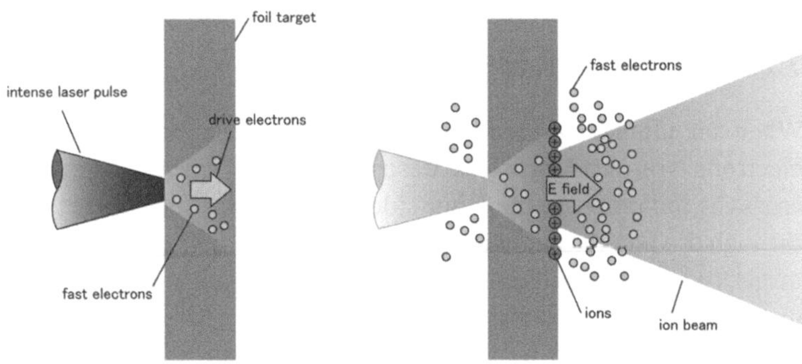

Figure 7.5 Schematic description of the TNSA model.

In general, TNSA proton beams have the following problems (of course there are numerous studies to improve these problems):

- The energy spectrum of the TNSA proton beam has Maxwellian form with a characteristic temperature (not monoenergetic beam) and with cut-off energy (maximum energy).
- So far, the maximum proton energy obtained with present laser facilities is several tens of mega-electron volts. It is required to intensify laser pulses for generating more energetic protons.
- The proton beam has large initial divergence.

Though laser-accelerated proton beams still face many problems in providing mega-electron-volt proton beams like those commonly used in particle accelerators, many studies have been carried out to resolve these issues.

7.3 Energy Increase in Proton Beam Acceleration for Ion Beam Analysis

To use the laser-accelerated proton beams for IBA, a stable supply of the proton beams over mega–electron volt energy is necessary. The maximum energy of the TNSA protons scales with the laser pulse energy [37]. The laser pulse with energy >10 J can produce a large number of protons over mega–electron volt energy, though the laser facilities that can deliver such pulses are large and their repetition rate is not so high (typically under several hertz). Therefore, it is required to increase the maximum energy of the proton beam by using the high-accessibility laser systems with lower energy (<1 J) and a high repetition rate (>10 Hz).

Here, we show efficient proton acceleration by the interaction of an intense femtosecond laser pulse with a solid foil. A disk-shaped aluminum coating (thickness 0.2 μm) on a polyethylene (PE) foil was irradiated at 2×10^{18} W/cm^2 intensity. The protons from the aluminum disk (diameter 150 μm to 15 mm) foil were accelerated to much higher energy in comparison with conventional targets such as PE and aluminum-coated PE foils. The fast electron signal along the foil surface was significantly higher than that from the aluminum-coated PE foil. The laser proton acceleration appears to be affected by the size of the surrounding conductive material.

7.3.1 Introduction

Since the first demonstration of laser proton acceleration, ion acceleration has been studied for enhancing proton energy much more for potential applications. Simply increasing the incident laser intensity [2, 38] enhances the proton energy. In the TNSA model, the strength of the electrostatic field generated at the rear surface of the target is proportional to the average energy (temperature) of the laser-accelerated electrons [13], and therefore the maximum energy of the protons is proportional to the electron temperature [14, 39]. The electron temperature is mainly proportional to the square of the laser intensity I in the range >10^{19} W/cm^2, and it has been actually observed that the maximum proton energy is proportional to $I^{0.5}$ [38]. The contrast ratio of the main laser pulse to the amplified spontaneous emission is also an important factor because it is deeply related to the energy transfer (absorptance) of the laser energy to the fast electrons. By controlling the plasma preformed by the amplified spontaneous emission (ASE) part in the pulse (it is called the "preplasma"), a 25% enhancement in the maximum energy is observed [40].

The target has been also structured. To enhance laser absorption, a single layer of polystyrene nanospheres is located at the front of the target surface, and then about 60% enhancement is observed in the maximum energy [41]. Most recently, by configuring the target to control the electric field for proton acceleration, about threefold enhancement of the maximum energy has been achieved [42, 43]. Buffechoux et al. show that the small foil target (200 μm × 200 μm × 0.2 μm) increases the maximum proton energy (threefold higher than a large foil) [42]. It is considered that trapping of the laser-accelerated electrons in the small target area enhances the electrostatic field for proton acceleration. However, the field is reduced not only by decelerated fast electrons (return current) but also by surrounding free electrons in the target foil. Accordingly, the target foil should ideally be electrically isolated [42, 43]. However, Toncian et al. [43] report that on the rear surface of a small isolated target, the sheath field can be reduced by blown-off electrons that flow around the target to the rear surface.

In this section, a configuration of the foil target is proposed to avoid the rapid sheath field reduction caused by circumjacent electrons streaming into the positively charged area formed by the laser pulse. The proposed target foil consists of a conductive material (aluminum) for generating electrons in abundance and an insulating material (PE) for decreasing electron flow to the rear surface. A small-diameter aluminum disk was deposited onto PE foil, and a laser pulse was focused onto the disk. With the present target, protons were accelerated to much higher energy in comparison with uncoated PE foil or PE foil completely coated with aluminum.

7.3.2 Experimental

The experiment was conducted using the T^6 laser system at the Institute for Chemical Research at Kyoto University. Laser pulses were generated from Ti:sapphire chirped pulse amplifiers. The pulse duration was 130 fs, and the central wavelength was 800 nm. The intensity contrast ratio at 200 ps prior to the pulse peak was 10^{-7} times the peak intensity. A laser pulse of 2×10^{18} W/cm^2 intensity, which corresponded to an energy of 40 mJ, was focused on the target foil to a spot size of 3 μm × 5 μm (full width at half maximum [FWHM]) by an F/3.2 off-axis parabolic mirror. Three types of foils were used in the experiments (Fig. 7.6). The first type was a 10 μm thick PE foil, which was expected to be a proton source containing abundant hydrogen atoms. The second type, a source of high-density electrons, was a 10 μm thick PE foil coated with a 0.2 μm layer of aluminum on the laser irradiation surface. The third type was a PE foil onto which an aluminum disk had been deposited. The aluminum disk was deposited on a 10 μm thick PE foil through a metal mask with a pinhole 150 μm to 15 mm in diameter. The aluminum disk was as thick as the aluminum layer on the second type of foil. Hereinafter, these foils are referred to as "PE foil," "Al-coated foil," and "Al disk foil," respectively. Each foil was tightened on a metal ring 8.8 cm in diameter. The ring was rotated to expose a fresh surface for each laser shot. The focal position for the foil surface was corrected before each laser shot by using a laser displacement sensor (Keyence, LK-G80), the accuracy of which was ±5 μm (less than the Rayleigh length of the focusing optics).

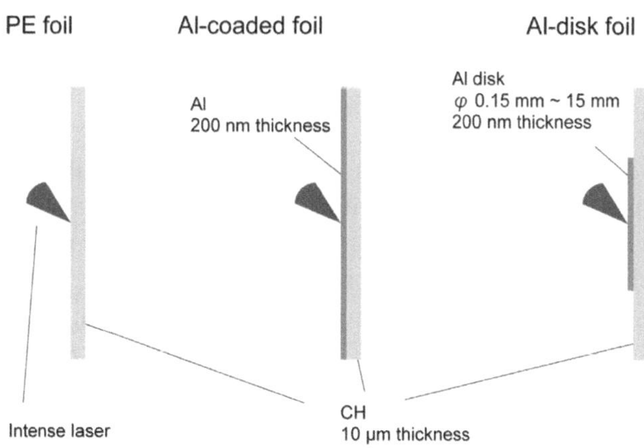

Figure 7.6 Schematic of three types of foil targets.

Behind the target foil, the proton energy spectra were measured by a time-of-flight (TOF) technique (Fig. 7.7). A plastic scintillator was set 2 m behind the laser-irradiated target. A pair of dipole magnets was set at the midstream part of the TOF tube to deflect electrons with energy less than 2 MeV. The spatial distribution of fast electron emission was also measured with imaging plates (IPs). The IPs were arranged at a distance of 80 mm behind the target over 180° horizontally on the laser incidence plane. The IPs were stacked in two layers. The second IP layer measured the distribution of electrons (>400 keV) filtered through 11 μm thick aluminum and the first IP layer.

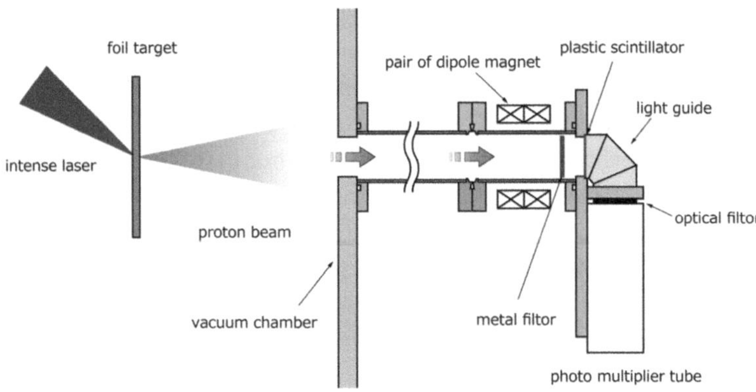

Figure 7.7 Schematic of a TOF experimental setup for measuring the proton energy spectrum.

7.3.3 Results and Discussions

The measured proton spectra are shown in Fig. 7.8 [44]. The maximum proton energy (maximum E_p) from the Al-coated foil is clearly lower than that from the PE foil. Higher-energy protons are emitted from the Al disk foil than from the other foils. The proton energy is influenced by the size of the surrounding metal region. In the energy range of 0.7–1 MeV, more protons are generated from the Al-coated foil than from the PE foil. Figure 7.9 shows the distribution of maximum E_p for different disk sizes of the aluminum coating [44]. The maximum E_p from the Al-coated foil is also plotted for the disk size of 8.8 cm, which corresponds to the foil size. The maximum E_p from the PE foil is also shown on the left edge of the graph. The maximum E_p from the Al disk foil gradually decreases as the aluminum disk diameter is increased beyond 2.5 mm.

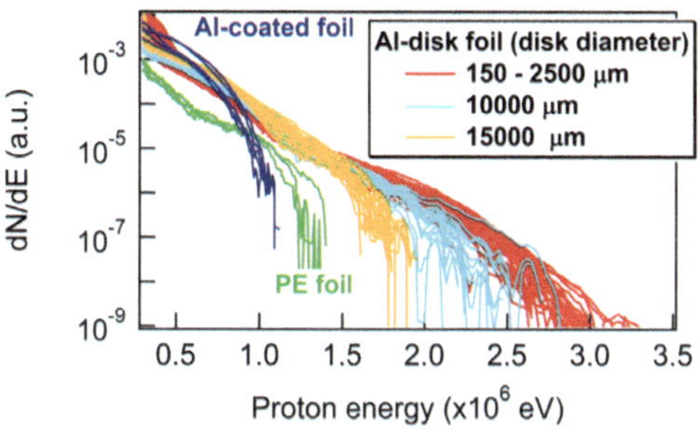

Figure 7.8 Proton spectra emitted from three types of targets. Each target was shot more than five times to confirm reproducibility. Reprinted from Ref. [44], with the permission of AIP Publishing.

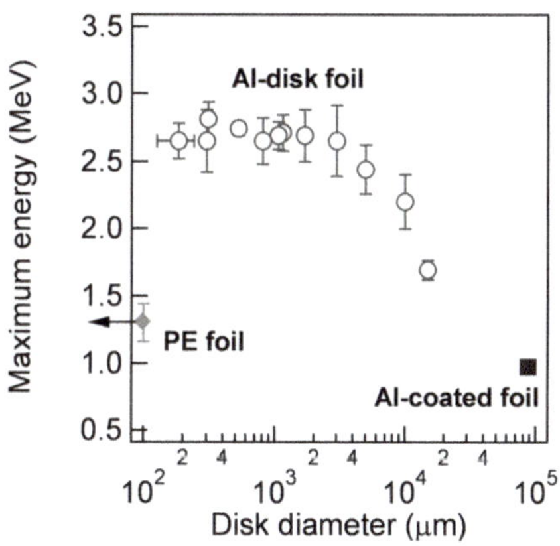

Figure 7.9 Maximum E_p distribution for deposited aluminum disks of various diameters. Maximum E_p is also plotted for a PE foil and an Al-coated foil. Reprinted from Ref. [44], with the permission of AIP Publishing.

The distribution of electrons with energy greater than 400 keV is shown in Fig. 7.10 [44]. Similar distributions of electrons are emitted from a PE foil and an Al disk foil. The yield of fast electrons from the Al disk foil is twofold higher than that from the PE foil. The distribution of electron emission is markedly different between the Al-coated foil and the other foils. In particular, the emission of electrons along the target foil surface is much higher for the Al-coated foil than for the other two foils. The distribution of electrons emitted from the Al-coated foil suggests that the time-averaged shape of the sheath electric field is clearly different.

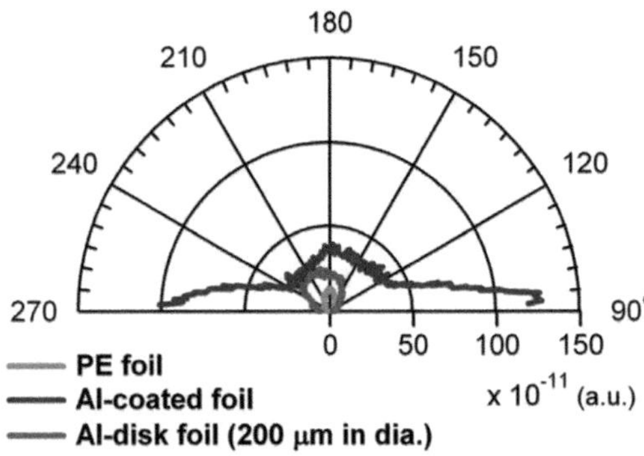

Figure 7.10 Electron distributions with energy greater than 400 keV were measured over 180° on the rear side of the target foil. The laser is directed from 350° in the coordinate system shown in the figure. Reprinted from Ref. [44], with the permission of AIP Publishing.

Differences in electron distribution and proton acceleration between the Al disk and Al-coated foils are considered to arise from the temporal evolution of the sheath electric field. Since the initial sheath electric fields on Al disk and Al-coated foils are considered to be the same, these differences in the experimental results occur through a relaxation process.

After laser pulse irradiation, the sheath field expands along the metal target at the speed of light [45]. The sheath potential propagates into the surrounding region, and the peak intensity decays rapidly. The expansion of the sheath field is limited by the size of the aluminum disk, and the sheath potential is maintained for longer duration than in the case of the Al-coated foil. The distribution of maximum E_p corresponding to the diameter of the aluminum disk also suggests the same trend in the expansion size of the sheath field.

Proton acceleration on the conductive foil is less effective than that on the insulating foil. The energy of protons emitted from the Al disk foil compared with that from the PE foil can be enhanced by increasing the efficiency of laser absorption and the generation of fast electrons in the metal coating. The experimental results on the electron distributions from these two targets indicate that the growth of the sheath fields is the same. The increase in the yield of 0.7–1 MeV protons from the Al-coated foil also suggests that the protons are generated from a larger area of the expanded sheath field. Compared with the Al disk and PE foils, the Al-coated foil produces a different electron distribution because of the expanded sheath field. The shape of the electric field on the Al-coated foil is distorted to become parallel to the target surface, owing to the expansion of the sheath field induced by the flow of free electrons. As a result, the lateral coulomb force is reduced and fast electrons are emitted in abundance. These results show that a target for efficient ion acceleration should generate a high yield of fast electrons to induce a strong sheath field and should be electrically isolated in order to maintain the field.

7.3.4 Summary

In summary, laser-accelerated proton energy is increased by using a PE foil with localized metal deposition. As the size of the deposited aluminum disk is increased, the maximum proton energy is reduced. The maximum proton energy is lower from the PE foil coated with aluminum over the entire surface, but the yield of low-energy protons is higher from a PE foil without metal coating. These results suggest that the sheath electric field expands around the metal foil and that the proton acceleration is strongly affected by the surrounding conductive material.

7.4 Application of Laser Accelerated Proton Beams for Ion Beam Analysis

7.4.1 Introduction

Since the first observations of the TNSA proton beam, a large number of studies have reported its applications, for example, in proton imaging for ultrafast phenomena [18, 19], creation of warm dense matter [20], cancer therapy using ion beams [21], and fast ignition for inertial confinement fusion [22]. There are a few reports of the application for a typical IBA, exemplified by Rutherford backscattering spectrometry, particle-induced X-ray emission, and nuclear reaction analysis. For example, Labaune et al. reported the measurement of a cross section of the $^{11}B(p, \alpha)2\alpha$ reaction in energetic laser plasma (an electron temperature of approximately 0.7 ± 0.15 keV) with laser TNSA proton beams [46]. However, the application of the TNSA proton for IBA has not been demonstrated. The considerable reasons are (a) a low repetition rate (low total integrated flux in comparison with the conventional particle accelerator), (b) stability of the TNSA beam, and (c) large radiation noise emitted from laser plasma (high-energy electrons, X-rays, or electromagnetic waves will damage a specimen and make it difficult to acquire data from diagnostic devices). Here, we describe a demonstration experiment for the application of laser-accelerated proton beams for IBA, with a compact and stable laser system. The TNSA proton beams are used for a $^{7}Li(p, \alpha)^{4}He$ reaction. A lithium fluoride (LiF) crystal is irradiated by the proton beams 500 times to detect the 8 MeV α particle from this reaction. To reduce large noise, a CR-39 nuclear track detector is used and significant signals generated by α particles have been observed.

7.4.2 Experimental

The experiment was carried out with a T6-laser system at the Institute for Chemical Research at Kyoto University [47, 48], which delivers 400 mJ 40 fs pulses at a center wavelength of 810 nm and a repetition rate of 5 Hz. The schematic of the experimental setup is shown in Fig. 7.11. The laser pulses for accelerating the TNSA proton beams were focused onto the PE film target (thickness 10 μm) by an F/3 off-axis parabolic mirror. The laser pulse was irradiated on the target at 45° from the target normal with a p-polarized geometry.

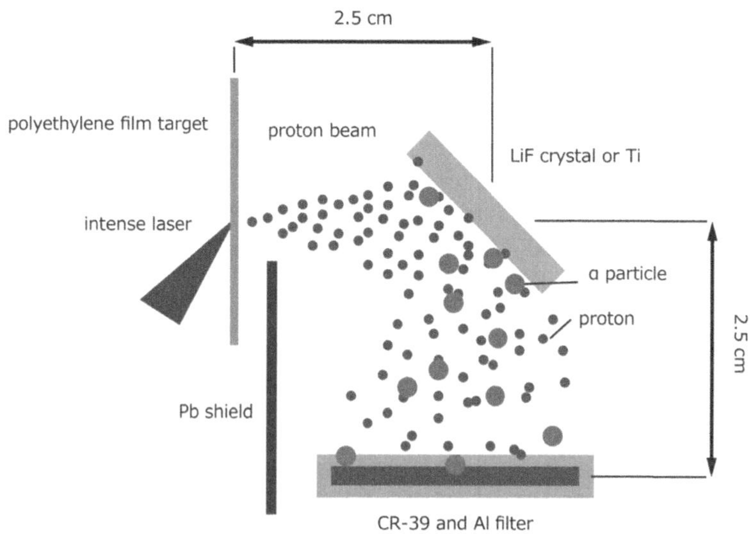

Figure 7.11 Schematic of the experimental setup.

Once the target film was irradiated with an intense laser pulse, the film was damaged and a large crater was formed around the irradiation position. To provide a fresh surface for each pulse with the repetition rate of 5 Hz, the target film was installed on a rotating stage. To irradiate laser pulses on the target foil precisely the rotating stage position was automatically corrected shot by shot. Figure 7.12b shows the picture of the target foil and a driving stage. The rotating stage was installed on a two-axis linear stage controlled by a computer. The displacement of the foil target surface in the laser direction at thc irradiation position was measured and recorded by using a laser displacement sensor before laser irradiation experiments, and the target holder was moved to correct the surface position in the laser direction for each laser irradiation. Figure 7.13 shows the target position displacement without the correction and after corrections with the target stages. The target position displacement was more than ± 15 μm in the direction of laser propagation without correction. After correction, the position displacement was extremely reduced, to be less than ± 1.2 μm in standard deviation.

TNSA proton beams were injected on the specimen. We used a LiF single-crystal plate (20 mm × 20 mm × 1 mm). The α particles from the ^{7}Li(p, α)^{4}He reaction in the LiF crystal were detected by CR-

39 nuclear track detectors. From the laser plasma produced by an intense laser pulse, numerous energetic electrons and photons were emitted. (Since the CR-39 has less sensitivity to them and has high sensitivity to ions, CR-39 can be placed at a 35 mm distance from the laser plasma and a 25 mm distance from the LiF crystal.) To estimate the number of scattering protons from the LiF crystal, a Ti plate (20 mm × 20 mm × 1 mm), which has a recoil cross section about 10 times larger than LiF, was also irradiated by the TNSA proton beams. To reduce the background ions, the CR-39 detectors were covered by an aluminum filter with a thickness of 22 μm, which can stop protons with energy <1.2 MeV though α particles with 8 MeV energy from the nuclear reaction can go through the Al filter.

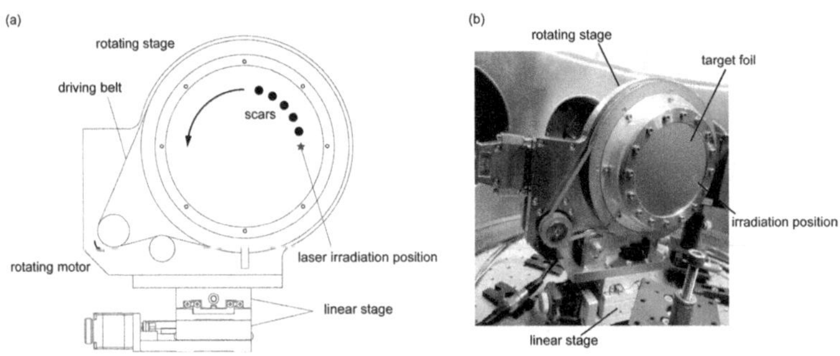

Figure 7.12 Schematic drawing of the target driving stage (a) and a picture of the target and the driving stage (b).

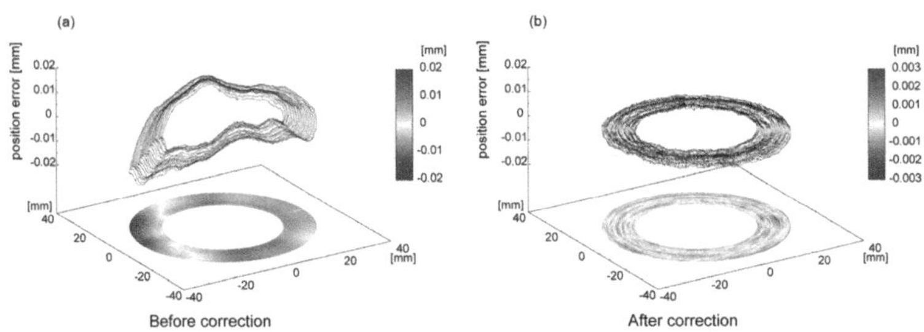

Figure 7.13 (a) The displacement of the foil surface in the laser direction at the laser irradiation position without corrections. The black lines show the positions of the foil surface, and the color scale shows its contour map. The position of the foil was changed by more than ±15 μm. (b) The displacement of the target foil surface after corrections. The displacement was reduced to less than ±3 μm, and the standard deviation was less than ±1.2 μm.

The typical energy spectrum and angular distributions of the TNSA proton beams for the laser focal-spot diameter and intensity of 3 μm × 5 μm at FWHM and 1×10^{19} W/cm^2, respectively, are shown in Fig. 7.14. The maximum energy of protons is about 2 MeV, with the divergence angle of about 35° at FWHM. The maximum energy is larger than the cut-off energy of the Al filter. Then, to reduce the maximum energy of protons, the intensity of the laser pulse was reduced by defocusing the laser pulse by placing the Al target at a 200 μm distance from the focal position. Although the cross section of ^{7}Li(p, α)^{4}He has a peak at the proton energy of 2 MeV [49], the background noise of the protons is large enough behind the signal of the α particle. The intensity of the laser pulse at a 200 μm distance is about 4×10^{17} W/cm^2.

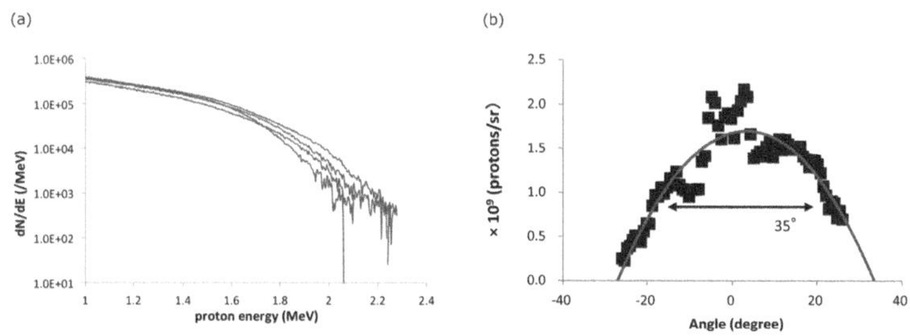

Figure 7.14 (a) Proton spectrum measured with a TOF energy spectrometer. The incident laser intensity is 2×10^{19} W/cm^2. The maximum energy of TNSA protons is around 2 MeV. (b) Angular distribution of the proton beam with energy of over 1.2 MeV measured with a CR-39 nuclear track detector covered with an Al filter. The divergence angle is 35° at FWHM.

7.4.3 Results and Discussions

Figure 7.15 shows a picture of CR-39 and a magnified image with an optical microscope to count the number of etch pits. The specimen is a LiF plate. The black dots in the microscope image show etch pits created by charged particles hitting the CR-39. The microscope images in a 1 cm × 1 cm area of CR-39 are acquired to count the number of etch pits for each specimen.

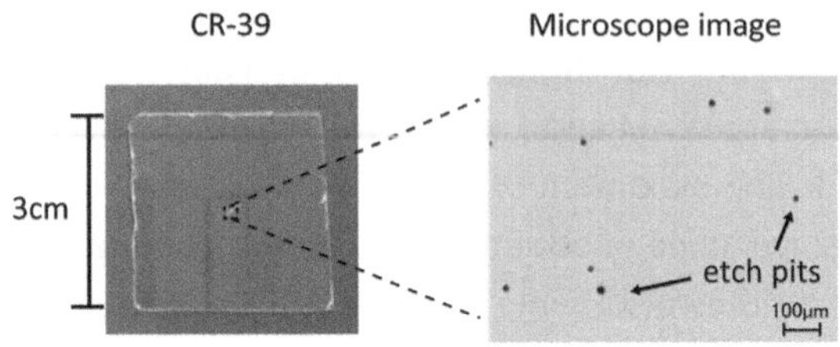

Figure 7.15 Picture of the CR-39 detector (left) and image obtained by an optical microscope (right). The black dots in the microscope image are etch pits created by ions hitting the CR-39.

Figure 7.16 shows the total number of etch pits. Each specimen was irradiated by 500 pulses of TNSA protons. The TNSA proton beams are accelerated by defocused laser pulses with the intensity of about 4×10^{17} W/cm^2. It is observed that 785 etch pits are created on the CR-39 for the LiF plate. In contrast, for the Ti plate and the absence of specimen, 136 and 165 etch pits are observed, respectively. This result suggests that the α particles from the ^{7}Li(p, α)^{4}He reaction in the LiF crystal are successfully detected.

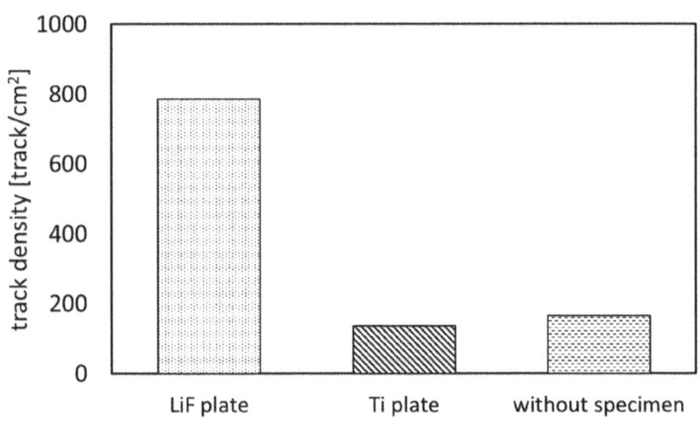

Figure 7.16 Counts of the etch pits for each specimen.

7.4.4 Summary

A demonstration experiment for the application of TNSA proton beams for IBA was performed. TNSA proton beams were injected on a LiF crystal plate to attempt to observe a ^{7}Li(p, α)^{4}He reaction with CR-39 nuclear track detectors. Etch pits were observed on the CR-39, and this observation is significant enough to suggest that the $^{7}Li(p, \alpha)^{4}He$ reaction was driven by TNSA proton beams.

The present experiment is still in its preliminary stage, and more detailed studies are necessary, but it is meaningful enough to demonstrate the potential of the TNSA proton beam with a tabletop laser for IBA to enliven future research.

References

1. Mangles, S. P., Murphy, C. D., Najmudin, Z., Thomas, A. G., Collier, J. L., Dangor, A. E., Divall, E. J., Foster, P. S., Gallacher, J. G., Hooker, C. J., Jaroszynski, D. A., Langley, A. J., Mori, W. B., Norreys, P. A., Tsung, F. S., Viskup, R., Walton, B. R. and Krushelnick, K. (2004). Monoenergetic beams of relativistic electrons from intense laser–plasma interactions, *Nature*, **431**, pp. 535–538.
2. Snavely, R. A., Key, M. H., Hatchett, S. P., Cowan, T. E., Roth, M., Phillips, T. W., Stoyer, M. A., Henry, E. A., Sangster, T. C., Singh, M. S., Wilks, S. C., MacKinnon, A., Offenberger, A., Pennington, D. M., Yasuike, K., Langdon, A. B., Lasinski, B. F., Johnson, J., Perry, M. D. and Campbell, E. M. (2000). Intense high-energy proton beams from Petawatt-laser irradiation of solids, *Phys. Rev. Lett.*, **85**, pp. 2945–2948.
3. Fukuda, Y., Faenov, A. Y., Tampo, M., Pikuz, T. A., Nakamura, T., Kando, M., Hayashi, Y., Yogo, A., Sakaki, H., Kameshima, T., Pirozhkov, A. S., Ogura, K., Mori, M., Esirkepov, T. Zh., Koga, J., Boldarev, A. S., Gasilov, V. A., Magunov, A. I., Yamauchi, T., Kodama, R., Bolton, P. R., Kato, Y., Tajima, T., Daido, H. and Bulanov, S. V. (2009). Energy increase in multi-MeV ion acceleration in the interaction of a short pulse laser with a cluster-gas target, *Phys. Rev. Lett.*, **103**, pp. 165002-1–165002-4.
4. Tabak, M., Hammer, J., Glinsky, M. E., Kruer, W. L., Wilks, S. C., Woodworth, J., Campbell, E. M., Perry, M. D. and Mason, R. J. (1994). Ignition and high gain with ultrapowerful lasers, *Phys. Plasmas*, **1**, pp. 1626–1634.
5. Atzeni, S. (1999). Inertial fusion fast ignitor: igniting pulse parameter window vs the penetration depth of the heating particles and the density of the precompressed fuel, *Phys. Plasmas*, **6**, pp. 3316–3326.
6. Tokita, S., Inoue, S., Masuno, S., Hashida, M. and Sakabe, S. (2009). Single-shot ultrafast electron diffraction with a laser-accelerated sub-MeV electron pulse, *Appl. Phys. Lett.*, **95**, pp. 111911-1–111911-3.

7. Tokita, S., Hashida, M., Inoue, S., Nishoji, T., Otani, K. and Sakabe, S. (2010). Single-shot femtosecond electron diffraction with laser-accelerated electrons: experimental demonstration of electron pulse compression, *Phys. Rev. Lett.*, **105**, pp. 215004-1–215004-4.

8. Rousse, A., Rischel, C., Fourmaux, S., Uschmann, I., Sebban, S., Grillon, G., Balcou, P., Förster, E., Geindre, J. P., Audebert, P., Gauthier, J. C. and Hulin, D. (2001). Non-thermal melting in semiconductors measured at femtosecond resolution, *Nature*, **410**, pp. 65–68.

9. Siders, W., Cavalleri, A., Sokolowski-Tinten, K., Tóth, Cs., Guo, T., Kammler, M., Horn von Hoegen, M., Wilson, K. R., von der Linde, D. and Barty, C. P. J. (1999). Detection of nonthermal melting by ultrafast X-ray diffraction, *Science*, **286**, pp. 1340–1342.

10. Ledingham, K. W. D., McKenna, P. and Singhal, R. P. (2003). Applications for nuclear phenomena generated by ultra-intense lasers, *Science*, **300**, pp. 1107–1111.

11. Remington, B. A., Arnett, D., Paul, R., Drake, and Takabe, H. (2003). Modeling astrophysical phenomena in the laboratory with intense lasers, *Science*, **384**, pp. 1488–1493.

12. Wilks, S. C., Langdon, A. B., Cowan, T. E., Roth, M., Singh, M., Hatchett, S., Key, M. H., Pennington, D., MacKinnon, A. and Snavely, R. A. (2001). Energetic proton generation in ultra-intense laser–solid interactions, *Phys. Plasmas*, **8**, pp. 542–549.

13. Hatchett, S. P., Brown, C. G., Cowan, T. E., Henry, E. A., Johnson, J. S., Key, M. H., Koch, J. A., Langdon, A. B., Lasinski, B. F., Lee, R. W., Mackinnon, A. J., Pennington, D. M., Perry, M. D., Phillips, T. W., Roth, M., Sangster, T. C., Singh, M. S., Snavely, R. A., Stoyer, M. A., Wilks, S. C. and Yasuike, K. (2000). Electron, photon, and ion beams from the relativistic interaction of Petawatt laser pulses with solid targets, *Phys. Plasmas*, **7**, pp. 2076–2082.

14. Kaluza, M., Schreiber, J., Santala, M. I. K., Tsakiris, G. D., Eidmann, K., Meyer-ter-Vehn, J. and Witte, K. J. (2004). Influence of the laser prepulse on proton acceleration in thin-foil experiments, *Phys. Rev. Lett.*, **93**, pp. 045003-1–045003-4.

15. Borghesi, M., Mackinnon, A. J., Campbell, D. H., Hicks, D. G., Kar, S., Patel, P. K., Price, D., Romagnani, L., Schiavi, A. and Willi, O. (2004). Multi-MeV proton source investigations in ultraintense laser-foil interactions, *Phys. Rev. Lett.*, **92**, pp. 055003-1–055003-4.

16. Strickland, D. and Mourou, G. (1985). Compression of amplified chirped optical pulses, *Opt. Commun.*, **55**, pp. 447–449.

17. Cowan, T. E., Fuchs, J., Ruhl, H., Kemp, A., Audebert, P., Roth, M., Stephens, R., Barton, I., Blazevic, A., Brambrink, E., Cobble, J., Fernández, J., Gauthier, J.-C., Geissel, M., Hegelich, M., Kaae, J., Karsch, S., Le Sage, G. P., Letzring, S., Manclossi, M., Meyroneinc, S., Newkirk, A., Pépin, H. and Renard-LeGalloudec, N. (2004). Ultralow emittance, multi-MeV proton beams from a laser virtual-cathode plasma accelerator, *Phys. Rev. Lett.*, **92**, pp. 204801-1–204801-4.

18. Romagnani, L., Fuchs, J., Borghesi, M., Antici, P., Audebert, P., Ceccherini, F., Cowan, T., Grismayer, T., Kar, S., Macchi, A., Mora, P., Pretzler, G., Schiavi, A., Toncian, T. and Willi, O. (2005). Dynamics of electric fields driving the laser acceleration of multi-MeV protons, *Phys. Rev. Lett.*, **95**, pp. 195001-1–195001-4.

19. Sokollik, T., Schnürer, M., Ter-Avetisyan, S., Nickles, P. V., Risse, E., Kalashnikov, M., Sandner, W., Priebe, G., Amin, M., Toncian, T., Willi, O. and Andreev, A. A. (2008). Transient electric fields in laser plasmas observed by proton streak deflectometry, *Appl. Phys. Lett.*, **92**, pp. 091503-1–091503-3.

20. Patel, P. K., Mackinnon, A. J., Key, M. H., Cowan, T. E., Foord, M. E., Allen, M., Price, D. F., Ruhl, H., Springer, P. T. and Stephens, R. (2003). Isochoric heating of solid-density matter with an ultrafast proton beam, *Phys. Rev. Lett.*, **91**, pp. 125004-1–125004-4.

21. Bulanov, S. V. and Khoroshkov, V. S. (2002). Feasibility of using laser ion accelerators in proton therapy, *Plasma Phys. Rep.*, **28**, pp. 453–456.

22. Roth, M., Cowan, T. E., Key, M. H., Hatchett, S. P., Brown, C., Fountain, W., Johnson, J., Pennington, D. M., Snavely, R. A., Wilks, S. C., Yasuike, K., Ruhl, H., Pegoraro, F., Bulanov, S. V., Campbell, E. M., Perry, M. D. and Powell, H. (2001). Fast ignition by intense laser-accelerated proton beams, *Phys. Rev. Lett.*, **86**, pp. 436–439.

23. Tajima, T. and Dawson, J. M. (1979). Laser electron accelerator, *Phys. Rev. Lett.*, **43**, pp. 267–270.

24. Key, M. H., Cable, M. D., Cowan, T. E., Estabrook, K. G., Hammel, B. A., Hatchett, S. P., Henry, E. A., Hinkel, D. E., Kilkenny, J. D., Koch, J. A., Kruer, W. L., Langdon, A. B., Lasinski, B. F., Lee, R. W., MacGowan, B. J., MacKinnon, A., Moody, J. D., Moran, M. J., Offenberger, A. A., Pennington, D. M., Perry, M. D., Phillips, T. J., Sangster, T. C., Singh, M. S., Stoyer, M. A., Tabak, M., Tietbohl, G. L., Tsukamoto, M., Wharton, K. and Wilks, S. C. (1998). Hot electron production and heating by hot electrons in fast ignitor research, *Phys. Plasmas*, **5**, pp. 1966–1972.

25. Roth, M., Blazevic, A., Geissel, M., Schlegel, T., Cowan, T. E., Allen, M., Gauthier, J.-C., Audebert, P., Fuchs, J., Meyer-ter-Vehn, J., Hegelich, M.,

Karsch, S. and Pukhov, A. (2002). Energetic ions generated by laser pulses: a detailed study on target properties, *Phys. Rev. ST Accel. Beams*, **5**, pp. 061301-1–061301-8.

26. Kmetec, J. D., Gordon, C. L., Macklin, J. J., Lemoff, B. E., Brown, G. S. and Harris, S. E. (1992). MeV X-ray generation with a femtosecond laser, *Phys. Rev. Lett.*, **68**, pp. 1527–1530.

27. Reich, Ch., Gibbon, P., Uschmann, I. and Förster, E. (2000). Yield optimization and time structure of femtosecond laser plasma kα sources, *Phys. Rev. Lett.*, **84**, pp. 4846–4849.

28. Norreys, P. A., Santala, M., Clark, E., Zepf, M., Watts, I., Beg, F. N., Krushelnick, K., Tatarakis, M., Dangor, A. E., Fang, X., Graham, P., McCanny, T., Singhal, R. P., Ledingham, K. W. D., Creswell, A., Sanderson, D. C. W., Magill, J., Machacek, A., Wark, J. S., Allott, R., Kennedy, B. and Neely, D. (1999). Observation of a highly directional γ-ray beam from ultrashort, ultraintense laser pulse interactions with solids, *Phys. Plasmas*, **6**, pp. 2150–2156.

29. Hamster, H., Sullivan, A., Gordon, S., White, W. and Falcone, R. W. (1993). Subpicosecond, electromagnetic pulses from intense laser-plasma interaction, *Phys. Rev. Lett.*, **71**, pp. 2725–2728.

30. Jin, Z., Chen, Z. L., Zhuo, H. B., Kon, A., Nakatsutsumi, M., Wang, H. B., Zhang, B. H., Gu, Y. Q., Wu, Y. C., Zhu, B., Wang, L., Yu, M. Y., Sheng, Z. M. and Kodama, R. (2011). Tunable radiation source by coupling laser-plasma-generated electrons to a periodic structure, *Phys. Rev. Lett.*, **107**, pp. 265003-1–265003-5.

31. Freidberg, J. P., Mitchell, R. W., Morse, R. L. and Rudsinski, L. I. (1972). Resonant absorption of laser light by plasma targets, *Phys. Rev. Lett.*, **28**, pp. 795–799.

32. Brunel, F. (1987). Not-so-resonant, resonant absorption, *Phys. Rev. Lett.*, **59**, pp. 52–55.

33. Tajima, T., Habs, D. and Mourou, G. A. (2010). Highest intensities, shortest pulses, *Opt. Photonik*, **5**, pp. 24–29.

34. Kruer, W. L. and Estabrook, K. (1985). J $\times$ B heating by very intense laser light, *Phys. Fluids*, **28**, pp. 430–432.

35. Davies, J. R. (2009). Laser absorption by overdense plasmas in the relativistic regime, *Plasma Phys. Control. Fusion*, **51**, pp. 014006-1–014006-21.

36. Bezzerides, B., Gitomer, S. J. and Forslound, D. W. (1980). Randomness, Maxwellian distributions, and resonance absorption, *Phys. Rev. Lett.*, **44**, pp. 651–654.

37. Gaillard, S. A., Kluge, T., Flippo, K. A., Bussmann, M., Gall, B., Lockard, T., Geissel, M., Offermann, D. T., Schollmeier, M., Sentoku, Y. and Cowan, T. E. (2011). Increased laser-accelerated proton energies via direct laser-light-pressure acceleration of electrons in microcone targets, *Phys. Plasmas*, **18**, pp. 056710-1–056710-11.

38. Borghesi, M., Fuchs, J., Bulanov, S. V., Mackinnon, A. J., Patel, P. K. and Roth, M. (2006). Fast ion generation by high-intensity laser irradiation of solid targets and applications, *Fusion Sci. Technol.*, **49**, pp. 412–439.

39. Mora, P. (2003). Plasma expansion into a vacuum, *Phys. Rev. Lett.*, **90**, pp. 185002-1–185002-4.

40. McKenna, P., Carroll, D. C., Lundh, O., Nurnberg, F., Markey, K., Bandyopadhyay, S., Batani, D., Evans, R. G., Jafer, R., Kar, S., Neely, D., Pepler, D., Quinn, M. N., Redaelli, R., Roth, M., Wahlstrom, C. G., Yuan, X. H. and Zepf, M. (2008). Effects of front surface plasma expansion on proton acceleration in ultraintense laser irradiation of foil targets, *Laser Part. Beams*, **26**, pp. 591–596.

41. Margarone, D., Klimo, O., Kim, I. J., Prokůpek, J., Limpouch, J., Jeong, T. M., Mocek, T., Pšikal, J., Kim, H. T., Proška, J., Nam, K. H., Stolcová, L., Choi, I. W., Lee, S. K., Sung, J. H., Yu, T. J. and Korn, G. (2012). Laser-driven proton acceleration enhancement by nanostructured foils, *Phys. Rev. Lett.*, **109**, pp. 234801-1–234801-5.

42. Buffechoux, S., Psikal, J., Nakatsutsumi, M., Romagnani, L., Andreev, A., Zeil, K., Amin, M., Antici, P., Burris-Mog, T., Compant-La-Fontaine, A., d'Humières, E., Fourmaux, S., Gaillard, S., Gobet, F., Hannachi, F., Kraft, S., Mancic, A., Plaisir, C., Sarri, G., Tarisien, M., Toncian, T., Schramm, U., Tampo, M., Audebert, P., Willi, O., Cowan, T. E., Pépin, H., Tikhonchuk, V., Borghesi, M. and Fuchs, J. (2010). Hot electrons transverse refluxing in ultraintense laser-solid interactions, *Phys. Rev. Lett.*, **105**, pp. 015005-1–015005-5.

43. Toncian, T., Swantusch, M., Toncian, M., Willi, O., Andreev, A. A. and Platonov, K. Y. (2011). Optimal proton acceleration from lateral limited foil sections and different laser pulse durations at relativistic intensity, *Phys. Plasmas*, **18**, pp. 043105-1–043105-7.

44. Otani, K., Tokita, S., Nishoji, T., Inoue, S., Hashida, M. and Sakabe, S. (2011). Efficient laser-proton acceleration from an insulating foil with an attached small metal disk, *Appl. Phys. Lett.*, **99**(16), p. 161501.

45. Quinn, K., Wilson, P. A., Cecchetti, C. A., Ramakrishna, B., Romagnani, L., Sarri, G., Lancia, L., Fuchs, J., Pipahl, A., Toncian, T., Willi, O., Clarke, R. J., Neely, D., Notley, M., Gallegos, P., Carroll, D. C., Quinn, M. N., Yuan, X. H., McKenna, P., Liseykina, T. V., Macchi, A. and Borghesi, M. (2009).

Laser-driven ultrafast field propagation on solid surfaces, *Phys. Rev. Lett.*, **102**, pp. 194801-1–194801-4.

46. Labaune, C., Baccou, C., Depierreux, S., Goyon, C., Loisel, G., Yahia, V. and Rafelski, J. (2013). Fusion reactions initiated by laser-accelerated particle beams in a laser-produced plasma, *Nat. Commun.*, **4**, pp. 2506-1–2506-6.
47. Tokita, S., Hashida, M., Masuno, S., Namba, S. and Sakabe, S. (2008). 0.3% energy stability, 100-millijoule-class, Ti:sapphire chirped-pulse eight-pass amplification system, *Opt. Express*, **16**, pp. 14875–14881.
48. Inoue, S., Maeda, K., Tokita, S., Mori, K., Teramoto, K., Hashida, M. and Sakabe, S. (2016). Single plasma mirror providing 10^4 contrast enhancement and 70% reflectivity for intense femtosecond lasers, *Appl. Opt.*, **55**, pp. 6435–6435.
49. Kang, M., Hu, Y., Hu, H. and Zhu, S. (2012). Cosmic rays during BBN as origin of lithium problem, *J. Cosmol. Astropart. Phys.*, **5**, pp. 011-1–011-23.

Chapter 8

Prospect of in situ Diagnostics of Lithium-Ion Batteries

Kunioki Mima,[a] Yuki Orikasa,[b] and Akiyoshi Yamazaki[c]

[a]*The Graduate School for the Creation of New Photonics Industries, 1955-1 Kurematsucho, Nishi Ward, Hamamatsu, Shizuoka Prefecture 431-1202, Japan*

[b]*College of Life Sciences, Ritsumeikan University, 1-1-1, Nojihigashi, Kusatsu, Shiga Prefecture 525-8577, Japan*

[c]*Faculty of Pure and Applied Sciences, University of Tsukuba, 1-1-1 Tennodai, Tsukuba, Ibaraki Prefecture 305-8577, Japan*

k.mima@gpi.ac.jp

8.1 Introduction

In this chapter, the present status and prospects of the in situ diagnostic techniques for Li-ion batteries (LIBs) are presented. To characterize LIBs, microscopic diagnostics for the lattice structure of active materials, nanoscale elemental distributions near the interfaces between active particles and electrolyte, and mesoscopic and macroscopic diagnostics for Li dynamics in the LIB's electrodes and cells are important. For these purposes, various diagnostic methods have been developed and applied [1, 2]. Electron

Lithium-Ion Batteries: Overview, Simulation, and Diagnostics
Edited by Yoshiaki Kato, Zenpachi Ogumi, and José Manuel Perlado Martin

ISBN 978-981-4800-40-2 (Hardback), 978-0-429-25934-0 (eBook)
www.panstanford.com

microscope [2, 3], X-ray diagnostics [4–6], atomic force microscope [7], neutron diagnostics [8–13], Raman spectroscopy [14–17], and ion beam analysis (IBA) [18–23] have been applied.

In Section 8.2, the status of these in situ measurements is summarized and as examples, microscopic measurements by means of X-ray, neutron diagnostics, and Raman spectroscopy are briefly reviewed. Since lithium behavior on a microscopic (say nanometer) scale is not directly related to the macroscopic performance of LIBs, the in situ measurement of Li distribution on the mesoscopic (10 nm to 100 μm) scale by neutron imaging and IBA would be of help to understand and to connect microscopic electrochemical processes to macroscopic LIB behavior.

As for in situ measurement with IBA [13], Section 8.3 describes the capability and limitation of the techniques to study the mesoscopic lithium distribution in LIB electrodes. Recently, time-dependent Li-ion and other elemental distributions have been measured by particle-induced γ-ray (PIGE) and particle-induced X-ray (PIXE) by using the proton microbeam at Takasaki Ion Accelerators for Advanced Radiation Application (TIARA), National Institutes for Quantum and Radiological Science and Technology (QST), Japan, where the 478 keV γ-ray from the $^7Li(p,p'\gamma)^7Li$ reaction was used to see Li distributions during charge and discharge. Finally, the potential application of nuclear reaction analysis (NRA) in the measurement of Li distributions without destroying the sample is discussed in Section 8.4.

8.2 Overview of Methods for in situ Diagnostics of Lithium-Ion Batteries

As described by the Department of Energy (DOE) report [1], the state-of-the-art characterization techniques have reached a high level of maturity in probing and monitoring the dynamic composition and structural changes associated with charge and mass transport, the detailed mechanism of the system operation, functionality of its components, and insights into the potential failure mechanism. Modern batteries contain a wide range of metallic, organic, and inorganic components with dimensions of interest ranging from Angstrom up to the centimeter range. Therefore, it is crucial to investigate various phenomena in different spatial scales by using suitable diagnostic techniques. Most of the presently used techniques are oriented to obtain information about Li behavior on the atomic scale and/or on the macroscopic scale. As an example, Li atoms in a crystal have been observed by using newly developed transmission electron microscope techniques [2].

In the following section, the typical diagnostics of LIB performance are described.

8.2.1 X-Ray Diagnostics

In the past, analysis of LIB performance was possible only for the stationary state or by computer simulations. Recently, for measuring nonstationary states, in situ diagnostic techniques were developed in X-ray diffraction (XRD) and X-ray absorption fine structure (XAFS) by using intense synchrotron radiation. In this section, overviewed are the in situ X-ray diagnostics of LIB performance.

In situ XRD and extended XAFS (EXAFS)/X-ray absorption near-edge structure (XANES) have been used to measure the phase transition in the charge and discharge states of LIBs made by $LiFePO_4$ (LFP). The measurements have been carried out by the RISING beam line at SPring-8 [5]. The time-dependent XRD spectra during the fast and slow charge/discharge reaction of the LFP battery have been successfully taken by Uchimoto et al. There are two stable phases, LFP and $FePO_4$ (FP). But the metastable state of Li_xFePO_4 appears in the fast reaction. The metastable state is encircled by a rectangle in Fig. 3.11, Chapter 3, in this book [5].

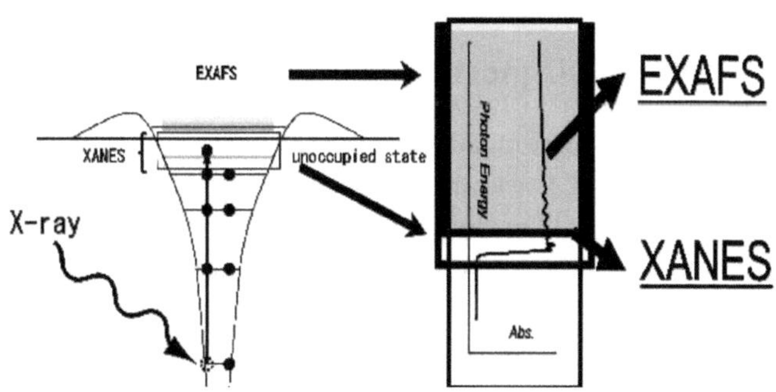

Figure 8.1 Electron excitation diagram of extended X-ray absorption fine structure (EXAFS) and X-ray absorption near-edge structure (XANES) measurement of a LIB cell with the SPring-8 (8 GeV Synchrotron Radiation). Reprinted with permission from Ref. [5]. Copyright (2013) American Chemical Society.

For the same sample, they measured the time-dependent spectra of XANES to obtain the change of the fraction of LFP versus FP, as shown in Fig. 8.2a. The XANES data are consistent with the fraction calculated by the charge transfer, as shown in Fig. 8.2b.

The above in situ techniques will be extended to get useful information on the performance of various LIBs. The details of the techniques on how to fabricate the cell and set up the diagnostic system are described in Refs. [5, 24].

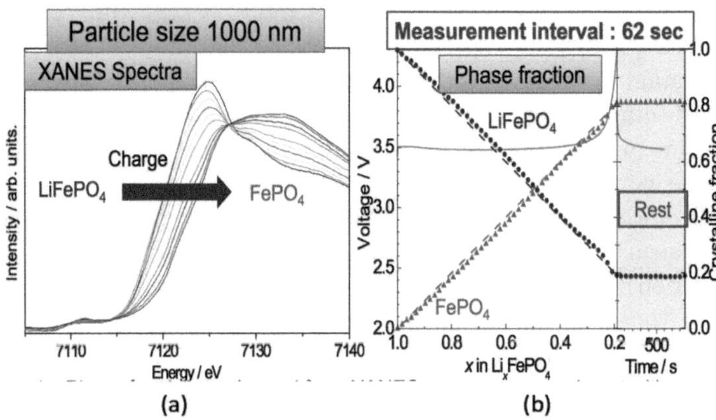

Figure 8.2 In situ measurement of a charging LFP electrode by XANES. (a) Absorption edge spectra changes corresponding to the transition from LFP to FP. (b) Fraction of LFP and FP calculated by charging current. Reprinted with permission from Ref. [24]. Copyright (2013) American Chemical Society.

8.2.2 Neutron Diagnostics

The setup of the in situ diagnostics of LIBs by using neutron beams is shown in the Fig. 8.3. For example, neutron diagnostics provides the depth distribution of Li in electrodes by the energy spectra of tritons produced by the following n-Li nuclear reaction (Eq. 8.1):

$$n + {}^{6}_{3}\text{Li} \rightarrow {}^{4}_{2}\text{He}(2055\ \text{keV}) + {}^{3}_{1}\text{H}(2727\ \text{keV}) \tag{8.1}$$

In this method, energy spectra of 2.06 MeV α-particles or 2.73 MeV triton are detected to see the depth profile of Li concentration. In other words, depth distribution is inferred by the energy loss of the α-particles or tritons when they are escaping from the sample.

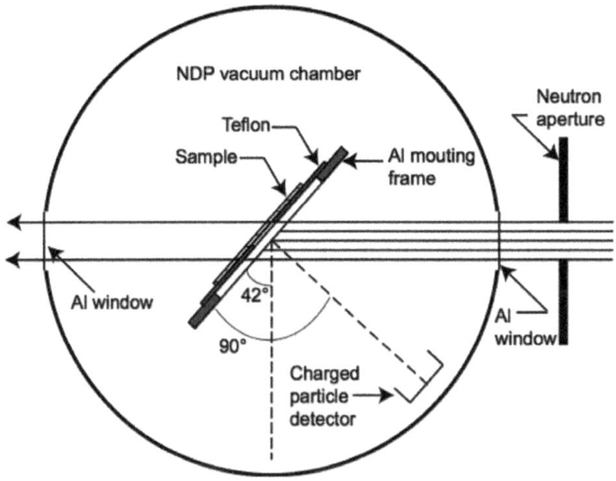

Figure 8.3 A schematic setup of the neutron diffraction for the in situ measurement of a Li-ion battery [8, 9]. Reprinted from Ref. [9], Copyright (2011), with permission from Elsevier.

An example of the neutron depth profiling (NDP) measurements is the Li distribution in aged commercial LIB cells [9]. A commercial LIB cell consists of a graphite anode and a cathode comprising LFP nanoparticles (40–50 nm). The anode is made of copper coated with graphite, and the cathode contains layers of LFP nanoparticles on an aluminum substrate. Long films containing anode and cathode with a separator in-between are rolled up and then packed into a cylindrical can. The electrolyte in the separator is lithium hexafluorophosphate ($LiPF_6$) salt in an organic solvent. The cell has a nominal capacity of 2.3 Ah, or 7.6 Wh. In aging cells, C1–C16, which are charged and discharged with 1C–16C, respectively, cycling of the cells was continued till the capacity was reduced to 80% of the initial capacity. The cells were cycled at 45°C. According to the automotive industry standard, a cell is assumed dead when its capacity is below 80% of the original one.

For the C0 cell, only one charge/discharge cycle at 1 C rate was established as the baseline cell. The effect of the charging or discharging current rate (C rate) was studied by the NDP on C0 (new cell) and C16—aged cell cycled from 60% to 70% state of charge (SOC) at 16 C rate.

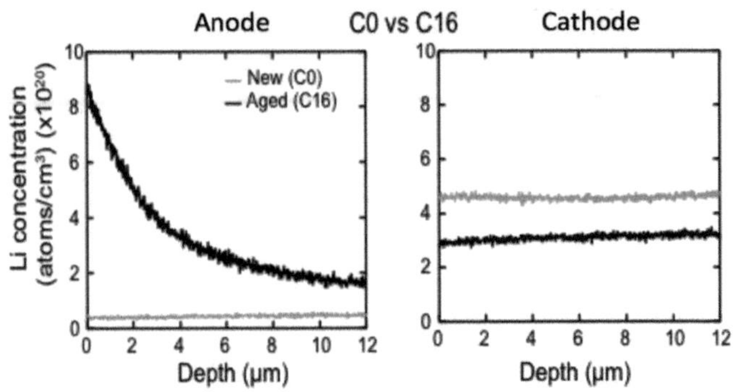

Figure 8.4 Depth profiles of Li concentration for the nominal sample C0 (light gray) and the aged sample C16 (black) at a high C rate and with a high state of charge (SOC). Reprinted from Ref. [9], Copyright (2011), with permission from Elsevier.

Figure 8.4 shows the results of the Li depth distribution measured by the NDP technique [9]. In the measurements, the LIB cells were made of 1.5 m long roll-up films. The samples were cut out from various locations of the roll-up film. Figure 8.4 is for the samples taken from the location close to the core, where the tritium spectra for the n-^{6}Li reaction were detected.

In those samples, the Li distributions are uniform in the cathode (positive electrode) but those in the anode (negative electrode) layer made by graphite are nonuniform for sample C16, which is the aged sample. The Li depth profiles in the positive electrode are uniform both for the new (C0) and aged (C16) samples, as shown in Fig. 8.4. But the Li concentration is lower for the aged sample in the state of charge of 60% $\approx$ 70%.

Neutron small-angle scattering has also been applied to study positive and negative electrode materials. The neutron small-angle scattering technique is efficient for obtaining the structures of compounds containing light elements such as lithium. Nagao et al. used the small and wide-angle neutron scattering technique to obtain long-range structural information and to clarify the local structure to confirm the lithiation mechanism [11].

Figure 8.5 shows a typical neutron small-angle scattering diagnostic system, where the measurable spatial scale d by the neutron small-angle scattering depends upon the neutron wavelength λ and the angular resolution $\Delta\theta$ of the detector. The maximum measurable spatial scale $\frac{d_{\max}}{\lambda}$ is inversely proportional to the scattering resolution $\Delta\theta$ according to Fourier inversion. The measurable scale extends typically from 0.1 nm to 10 nm.

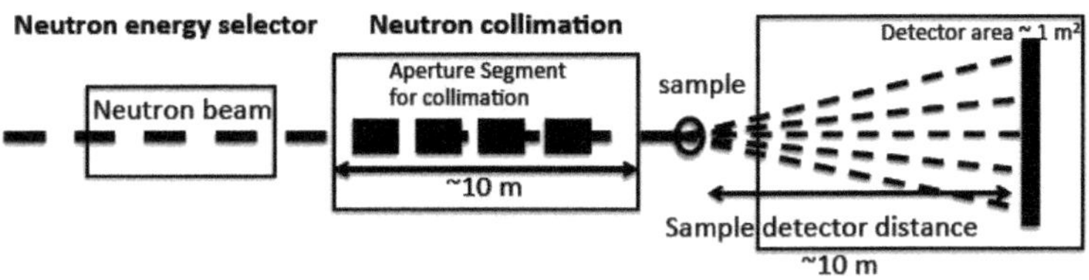

Figure 8.5 Schematic view of a typical neutron small-angle scattering system [13].

8.3 In situ Analysis by Raman Spectroscopy

The recent advances in in situ Raman spectroscopy for studying LIBs are reviewed by Stancovski et al. and Novak et al. [14, 15]. During a relatively short period of time (1995–2013), spectroscopic electrochemical techniques were developed for characterizing LIBs.

In in situ Raman spectroscopy, a laser irradiates a sample electrode in a test cell specially designed as shown in Fig. 8.6. An example of Raman spectra of the active material $LiCoO_2$ is shown in Fig. 8.7, which indicates that the $LiCoO_2$ cathode is degraded by forming Co_3O_4 when Li is deintercalated. The Raman shift at 674 cm^{-1} corresponds to the production of Co_3O_4.

The applied voltage dependence of the Raman shift spectra of a cathode made by $LiMn_2O_4$ has been found by Ito et al. [16]. Figure 8.8a shows the Raman spectra change at 4.1 V. This indicates the transition of the crystal structure of $Li_xMn_2O_4$ ($1 > x > 0.5$) when the deintercalation proceeds. On the other hand, Fig. 8.8b indicates that when Li is intercalated into $Li_xMn_2O_4$ ($1 > x > 0.5$), the Raman shift spectrum changes inversely. This means that the crystal structure of $Li_xMn_2O_4$ is inversely changed at 4.1 V. As shown here, the molecular processes in the electrochemical reactions of LIB electrodes can be detected by in situ Raman spectroscopy.

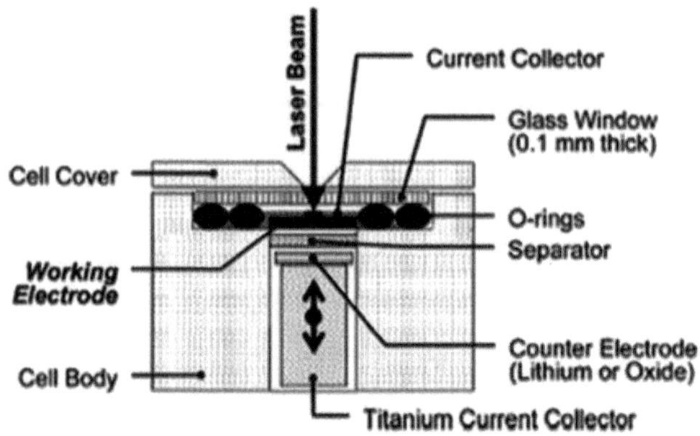

Figure 8.6 Experimental setup of Raman spectroscopy for the in situ measurement of a lithium-ion battery. Reprinted from Ref. [14], Copyright (2013), with permission from Springer Nature.

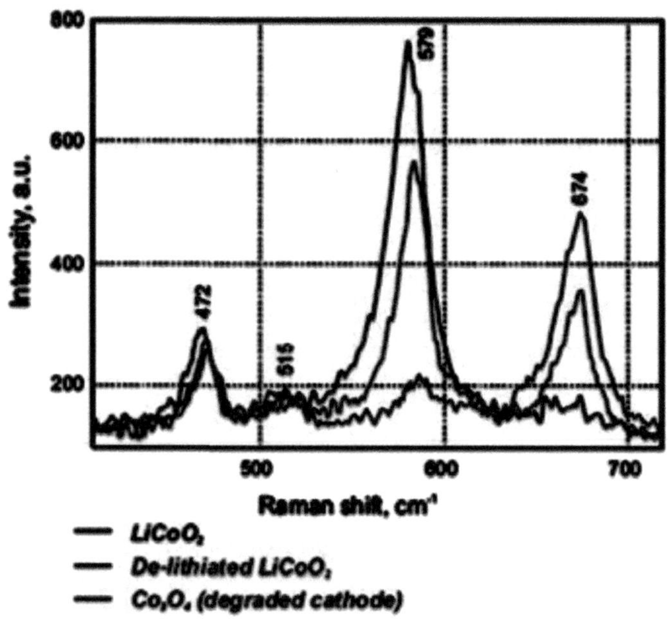

Figure 8.7 Raman shifts for a fully discharged cathode ($LiCoO_2$), a partially charged cathode, and a degraded cathode (Co_3O_4). Reprinted from Ref. [14], Copyright (2013), with permission from Springer Nature.

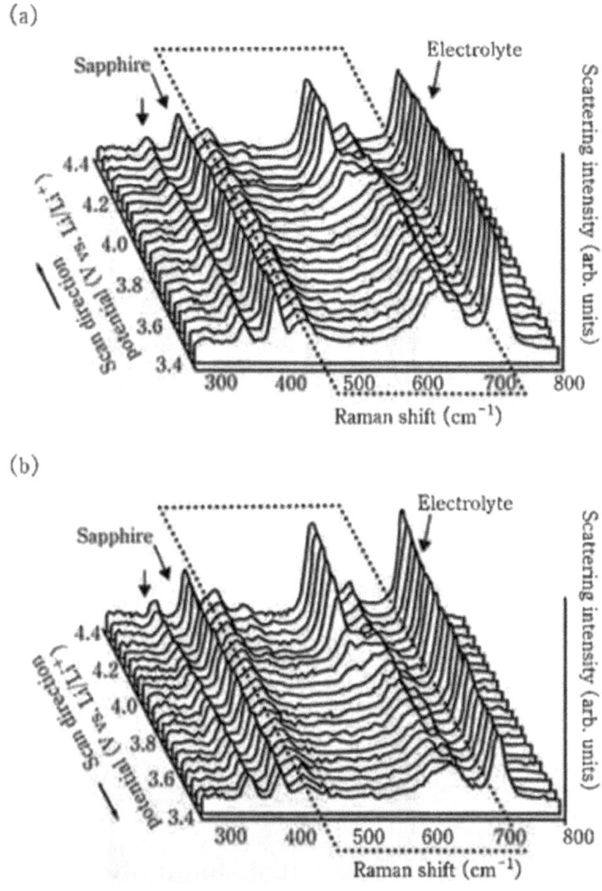

Figure 8.8 In situ Raman spectra of a $Li_xMn_2O_4$ cathode. (a) The voltage is scanned positively, (b) and the voltage is scanned negatively. Reprinted from Ref. [16], Copyright (2003), with permission from Elsevier.

8.4 In situ Measurement by Ion Beam Analysis

8.4.1 Introduction

IBA [22] is a powerful tool for quantitatively characterizing elemental concentrations. Furthermore, micro ion beam scanning technology allows us to study time-dependent elemental distributions with a high spatial resolution during charge or discharge. PIXE and PIGE techniques have been widely used for material study. Micro-PIXE and micro-PIGE technologies have been utilized for measuring lithium concentration distribution [17–22].

Mesoscopic elemental distributions in working LIBs without disassembling them have been measured recently using in situ IBA. Since the dynamics of lithium ions are crucial for the performance of LIBs, it is important to measure the Li concentration distribution in a working LIB. For in situ IBA of LIBs, it is necessary to inject ion beams into the electrode through a metal-laminated plastic film by which a battery is tightly sealed. Since the ion beams for PIXE, PIGE, and NRA measurements are scattered and lose their energy significantly in the sealed film, it is necessary to design a special test cell that has an ion beam injection window through which the ion beam energy loss is small. A light-ion microbeam line and its scanning system have been developed at TIARA, QST, which is the center of excellence for the ion beam technology and its application in Japan [23]. By combining the microbeam line with PIXE, PIGE, and NRA techniques, we can obtain 2D maps of various elements over a wide range of atomic numbers [24]. In this section, we report the capability of the in situ measurement of the ^{7}Li spatial distribution in a LIB positive electrode in a special LIB test cell. Here, we used the micro-PIXE and micro-PIGE techniques related to ^{7}Li(p,p') inelastic scattering and NRA. Direct observation of the temporal evolution of the lithium-ion distribution in the positive electrode is reported.

8.4.2 PIXE and PIGE Experiments for In situ Diagnostics

8.4.2.1 Sample fabrication

A LIB test cell has been fabrication for in situ measurements. The photograph and schematic illustrations of the test cell are shown in Fig. 8.9. The case of the cell is made of Teflon 40 mm in width, 40 mm in height, and 20 mm in thickness. It has a cylindrical cavity 5.5 mm in diameter at the center. The cavity, filled with an electrolyte (1 M $LiPF_6$/EC-EMC 3:7 mol), is sealed by polyether ether ketone (PEEK) of 2 mm thickness on one side and stainless steel of 3 mm thickness on another side. The PEEK plate has a circular hole 1 mm in diameter for ion beam irradiation. This hole is sealed by a 7.5-μm-thick Kapton film for preventing the electrolyte from flowing out of the cavity.

The positive electrode is composed of LFP as the active material, carbon black as the conductive material, and polyvinylidene fluoride as the binder in a ratio of 80:10:10 by weight. It is deposited on a thin aluminum foil used as a current collector. The thickness of the positive electrode material is approximately 20 μm. The top surface of the electrode material is covered by a 7.5-μm-thick Kapton film to prevent the electrode material surface from being soaked directly by the electrolyte. This positive electrode assembly is inserted into the cavity and placed just next to the packing Kapton film on the PEEK plate.

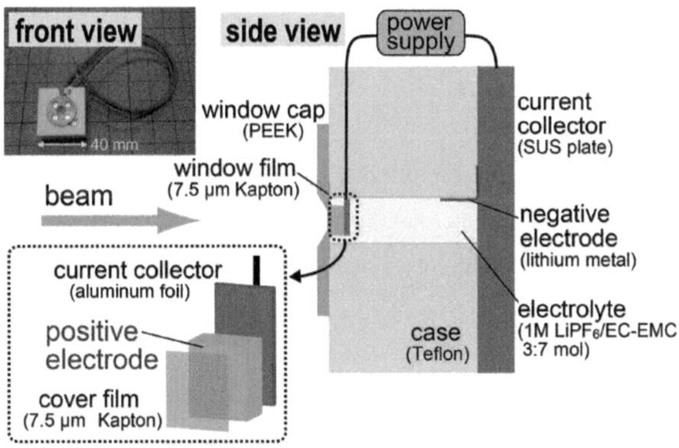

Figure 8.9 Photograph (front view, upper left) and schematic illustrations (side view) of a lithium-ion battery test cell for the in situ measurement of the spatial distribution of lithium in a LIB positive electrode. Reprinted from Ref. [22], Copyright (2016), with permission from Elsevier.

As the negative electrode, a thin lithium metal foil is inserted into the cavity and soaked in the electrolyte by placing the foil between the Teflon body and a stainless-steel plate current collector. The aluminum foil and the stainless-steel plate are connected to a potentiostat for charging the cell. The charging protocol, including the charging conditions and the measuring procedure, will be described in the next section.

8.4.2.2 Experimental procedure

The measurements have been carried out at TIARA, of QST, with a 3 MeV proton microbeam from a single-ended electrostatic accelerator. The diameter and current of the proton microbeam are typically 1.5 μm and 300 pA, respectively. A schematic illustration of the experimental setup for measurements of the X-rays and γ-rays is shown in Fig. 8.10. The proton microbeam in the vacuum chamber is injected into the air through a circular window covered with a 7.5-μm-thick Kapton film. Then, we scan the proton microbeam over the area of 800 μm × 800 μm to observe the evolution of the spatial lithium distribution in the positive electrode. The test cell is placed close to the beam extraction window by manipulation of a three-axes translation stage in the air, as shown in Fig. 8.10. The distance from the extraction window to the positive electrode sample is approximately 3 mm.

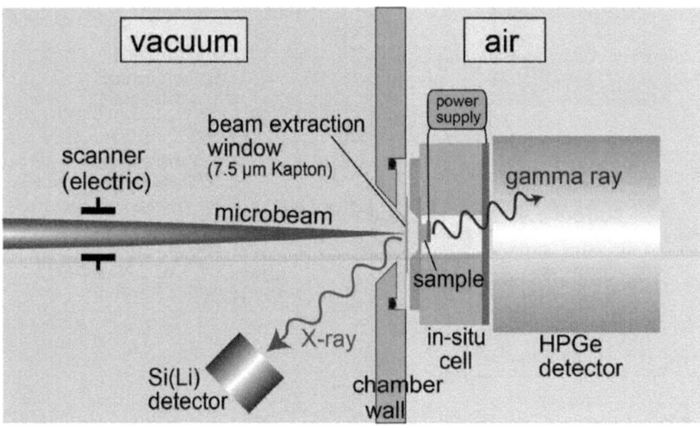

Figure 8.10 Schematic illustration of the liquid electrolyte LiB's experimental setup for PIXE and PIGE simultaneous measurement at QST. Reprinted from Ref. [22], Copyright (2018) with permission from Elsevier.

By taking into account the propagation through air and the three 7.5 μm Kapton films as described above, the energy deposition of the proton on the positive electrode is calculated with the simulation code SRIM2008 [25]. The incident proton energy at the surface of the positive electrode is estimated to be 2.6 MeV.

For X-ray detection, a Si(Li) detector is used, which is set in a vacuum chamber, as shown in Fig. 8.10. It has a Si(Li) crystal active area of 30 mm^2, and the distance from the crystal to the sample electrode is 24 mm. The lithium distribution is observed by micro-PIGE with the 478 keV γ-rays of the $^7Li(p,p')$ inelastic scattering. A high-purity germanium (HPGe) detector is used, which is set just behind the test cell, as shown in Fig. 8.10. The volume of the germanium crystal is 100 cm^3. A schematic illustration of the irradiating window of the test cell and a close-up photograph are shown in Fig. 8.11. In the experiment, the positive electrode is on the left in the beam irradiation window. The lithium distribution at the boundary of the positive electrode and the electrolyte can be observed. The scan area of the proton microbeam covering this boundary is shown by the dotted square in Fig. 8.11.

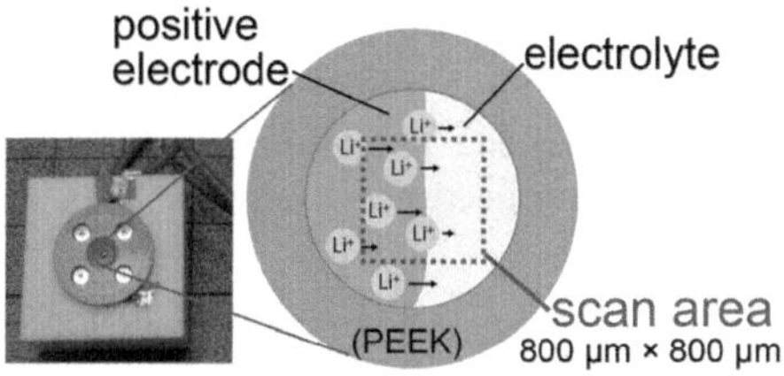

Figure 8.11 Schematic illustration of the irradiating window of the in situ test cell. The positive electrode is fixed on the left in the window circle. The beam scanning area is shown by a dotted square.

The test cell is charged with a constant current power supply by connecting the aluminum foil and the stainless-steel plate of the test cell with a constant current mode of 20 μA. The PIXE and PIGE images are recorded in sequence at the total charging times of 0 (initial condition), 50, 90, and 150 min., respectively. The charging was stopped to take images of each charging state for 20 min. to irradiate ion beams. We checked that the lithium distribution in LFP does not change during the beam irradiation time (in 20 min.).

8.4.2.3 Results and discussion

Typical X-ray and γ-ray energy spectra are shown in Figs. 8.12a and 8.12b, respectively. These spectra are for the positive electrode before charging. The characteristic X-rays of iron and the γ-rays of lithium in the positive electrode are the high peaks in Figs. 8.12a and 8.12b. In addition, emissions from some other elements are seen in the X-ray spectrum, in which several peaks are not from the positive electrode. They are the argon line from the air and the phosphorus and calcium lines from the calcium phosphate contained in the Kapton film.

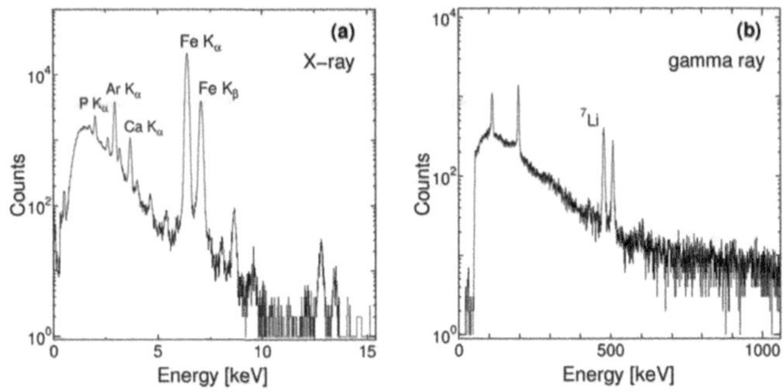

Figure 8.12 Typical X-ray (a) and γ-ray (b) energy spectra observed with a Si(Li) and an HPGe detector, respectively.

Concentration maps of the lithium and iron contained in the positive electrode at the charging times of 0, 50, 90, and 150 min. are shown in Fig. 8.13, where the lithium distribution is on the left (labeled Li) and the iron distribution in the center column (labeled Fe). These maps show that the yield of the lithium events gradually decreases with increasing charging time. Since we cannot determine the total amount of irradiating proton number in this experimental setup, the yield of the lithium event is normalized by the iron event pixel by pixel because the iron in the positive electrode active material is fixed during charge and discharge.

Normalized lithium maps are shown in the column on the right (labeled Li/Fe) in Fig. 8.13. They show that the lithium concentration near the edge of the electrode gradually decreases as the total charging time increases. The scan area is 800 μm × 800 μm for each map. Solid lines in the Li/Fe maps are the same as those shown in the maps of Fe distribution and indicate the approximate position of the positive electrode. In these images the γ-rays and the X-rays are not observed in the lower part and lower-left corner, because in these areas incident protons were stopped by the edge of the beam irradiation window.

These maps in Fig. 8.13 are at charging times of 0, 50, 90, and 150 min. We find that the lithium depression area appears at first near the edge and gradually spreads to the far side from the edge. These maps provide valuable information on lithium motion in the positive electrode of the working LIB. So we conclude that the combination of PIGE and PIXE is useful for the in situ measurement of the lithium concentration distribution.

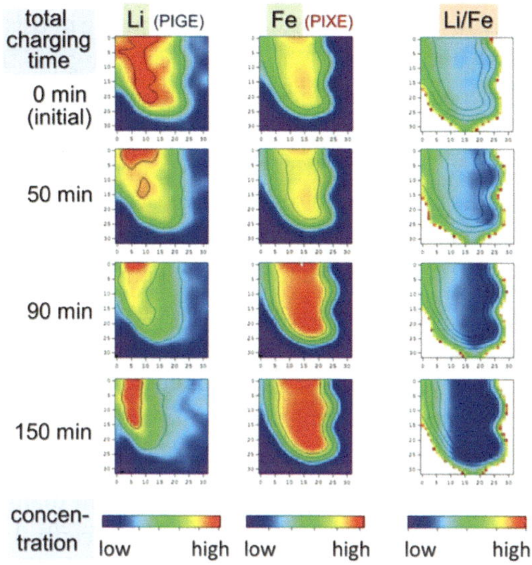

Figure 8.13 Spatial distributions of Li, Fe, and their ratio (Li/Fe) during the charge (0–150 min.).

8.4.3 Application in All-Solid-State LIBs

Recently explored is the applicability of in situ micro-PIXE and micro-PIGE in characterizing all-solid-state LIBs (ASSLIBs). The application of the in situ IBA in ASSLIB is relatively easier than that in liquid electrolyte LIBs since the solid-state electrolytes may be more tolerant to beam damage than the liquid electrolyte and also the vapor pressure of the electrolyte on the vacuum shielding film is not the problem. As proof-of-proof (POP) experiments of the applicability, three kinds of ASSLIBs have been tested. They are the Nb_2O_5 + LiAlGeP (LAGP) sample provided by the Battery Research Division of Toyota Motor Company; $LiCoO_2$ (LCO) + $Li_7La_3Zr_2O_{12}$ (LLZ) sample provided by IEK-1, Fz., Jülich, Germany; and the La-sulfate (LGPS) + LCO sample provided by the School of Energy Engineering, Tokyo Institute of Technology, Japan.

A test sample containing Nb_2O_5 + LAGP as shown in Fig. 8.14a is irradiated by a proton microbeam. The Nb_2O_5 secondary particles in LAGP is observed as vacant areas of Li, as shown in Fig. 8.14b.

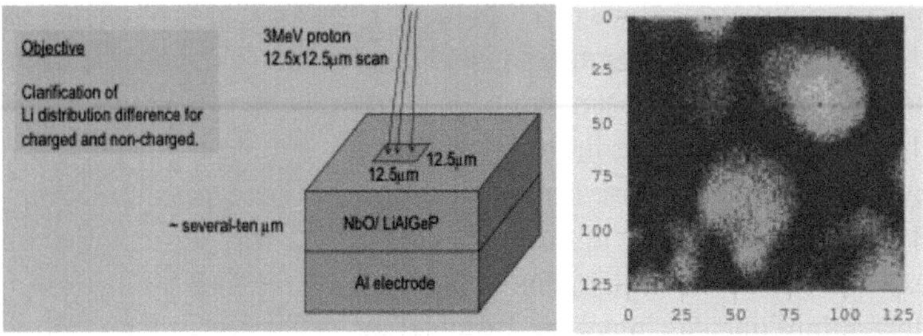

(a) (b)

Figure 8.14 (a) ASSLIB sample and PIGE/PIXE irradiation setup. (b) A micro-PIGE image of Nb_2O_5 secondary particles in LAGP.

As for the LCO + LLZ sample, the cubic area of the cross section of the test cell in Fig. 8.15 is placed at the 1 mm diameter central hole. Then the proton beam is injected from the central hole to take the images of Li (PIGE) and Zr (PIXE) and to see the change of the relative concentration of Li/Zr during discharge of charge. To shield the cross section from air, we cover the surface by a 12.5-μm-thick Kapton film. The change of the Li concentration between charge and discharge of this sample has been measured. The proof-of-principle in situ measurement of the working ASSLIB is under investigation.

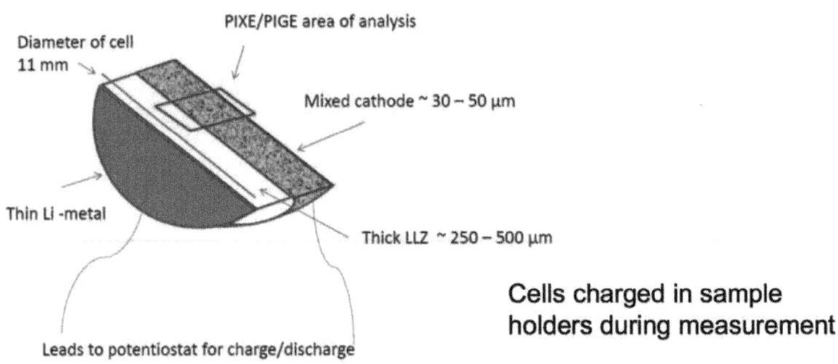

Figure 8.15 Test cell of Li metal/(LCO + LLZ)/LLZ.

8.4.4 Summary of Applications of PIGE and PIXE

We have developed a new method for in situ measurement of lithium distributions in LIBs using the micro-IBA technique. We have fabricated a special LIB test cell for the in situ measurement of elemental distributions in a LIB electrode and measured the time-dependent lithium-ion distribution during LIB charging. We have obtained 2D maps of lithium distributions and iron distributions by the micro-PIGE method and the micro-PIXE method, respectively. Distributions of lithium concentration have been determined by normalizing the lithium yield by the iron yield pixel by pixel. Decrease in the lithium concentration from the edge of the positive electrode during the charging has been clearly observed. Thus we have demonstrated for the first time the feasibility of the in situ measurement of a LIB by using the micro-IBA technology. This new technology is expected to provide a new diagnosing scheme for improving the performance of LIBs.

8.5 Capabilities of Rutherford Backscattering Spectroscopy and Nuclear Reaction Analysis

8.5.1 Introduction

Two more ways of in situ micro-IBA measurement of the Li depth distribution are Rutherford backscattering spectroscopy (RBS) and NRA [18]. The energy spectra of particles produced by ion beam nuclear interactions with samples allow obtaining information about Li transport properties related to LIB performance. The Li depth distribution measurement by RBS and/or NRA is advantageous since this characterization could be applicable to in situ measurements of 3D elemental distributions.

It is important to observe the nonuniformity of energy storage among the active particles. Nonuniformity arises during Li intercalation/deintercalation processes in active particles. Indeed, the differences in the local reaction rate on the mesoscopic scale may be due to the fact that not all active particles in the electrode effectively contribute in the same way to the electrochemical reaction during charge and discharge. Therefore, to connect the macroscopic and mesoscopic scales, it is crucial to know whether each active particle is contributing or not contributing to the reaction. So, it is necessary to find the "participation rate" of each secondary particle, which so far has not been characterized. In principle, combining the NRA with microbeam scanning procedures will allow the full

3D characterization of Li distribution in LIBs. However, because of the low-reaction cross section of the proton beam, we take spatial average of local (p,p) nuclear interaction data to get the quasi-3D depth distribution in the in situ measurements as discussed in the following section.

Several nuclear reactions can be chosen to characterize the Li depth profiling. These are $^{7}Li(^{3}He,\alpha_0)^{6}Li$, $^{7}Li(^{3}He,d)^{8}Be$, $^{7}Li(^{3}He,p)^{9}Be$, $^{7}Li(p,p_0)^{7}Li$, and $^{7}Li(p,\alpha_0)^{4}He$. However, as reported by Sagara et al. [26], among them, $^{7}Li(p,\alpha_0)^{4}He$ is the most suitable one to carry out the analysis because of the following reasons: (i) The intensity of the signal is proportional to the amount of the isotope in the sample, since the ^{7}Li isotope is the most abundant in natural Li; (ii) because of its large Q value (17.346 MeV) the signal-to-noise ratio is high and thick samples can be investigated; (iii) the reaction cross section has a broad maximum around the proton energy of 3 MeV, which allows one to estimate the depth profile directly from the energy spectrum of the emitted α-particles.

The Li depth profile was characterized by using the $^{7}Li(p,\alpha)^{4}He$ nuclear reaction with a proton beam impinging onto the sample surface at normal incidence.

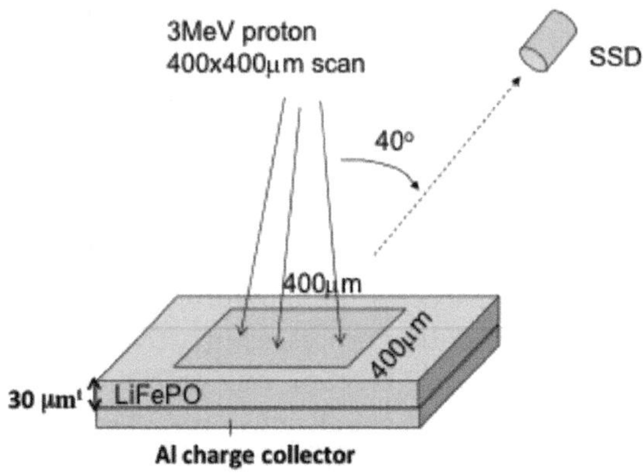

Figure 8.16 Ion beam analysis of LIBs setup.

The α-particles produced in the nuclear reaction and the protons backscattered by heavy elements (Fe, Mn, Ni, P, F, O, C) in the electrodes are detected by a Si-barrier detector as shown in Fig. 8.16, where an ion microbeam scans the surface area of 400 μm^2 to measure the 30-μm-thick active material layer (LFP layer) to obtain the spectra of protons due to RBS and α-particles due to NRA.

An example of the spectrum of the particles from the sample is shown in Fig. 8.17. The sample is a 30-μm-thick $LiFPO_4$ mixture with carbon and binder on the Al substrate. Here the spectra below 3 MeV are for protons due to RBS and those higher than 5 MeV are α-particle spectra. The signals between 3 MeV and 5 MeV are the pile-up signals that appear because of the high signal rate. The pile-up signals should not overlap with the spectral region of the α-particle signals.

The spectral shape of the α-particles of Fig. 8.17 is related to the Li depth distribution. The processes of analyzing data for the Li depth profiling are described in Section 5.2.

Note here that samples cannot be exposed to a beam directly when the samples are not compatible with vacuum and/or with air. In most cases, the sample surface contains materials like Li and liquid electrolytes or solid-state electrolyte evaporating gases or reacting with air. Therefore, the samples should be shielded or isolated from the vacuum chamber environment or air. So, we irradiate a sample with a beam through a several-micrometer-thick or thicker film in the in situ measurements. In this section, overviewed are the present status of the in situ IBS and the critical issues related to sample fabrication, experimental setup, and the POP experiments of the in situ IBA.

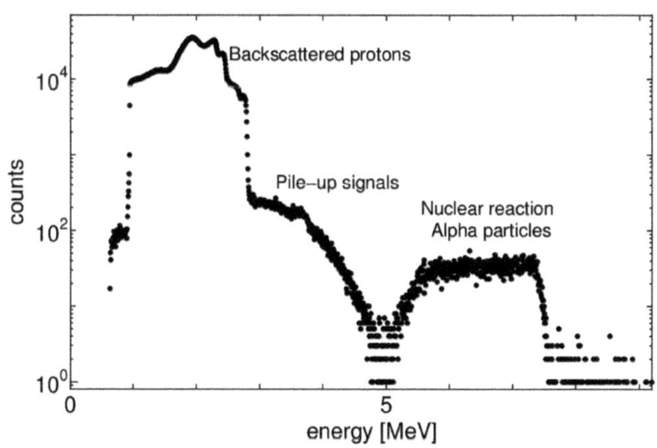

Figure 8.17 RBS and NRA spectra of the LFP cathode.

8.5.2 Present Status for Applying Nuclear Reaction Analysis to Li Depth Profiling

The actual NRA for the POP experiments toward in situ measurements [21] are presented in this section. The depth distributions of Li in LFP, $LiMnO_2$, or $LiNiO_2$ (LNO) positive electrodes provided by the Graduate School of Human and Environmental Studies at Kyoto University and the Toyota Central Research Laboratories are measured by NRA. The electrodes are fabricated by mixing the active material (LFP) with artificial carbon and binder (PVdF). The secondary active particle is $0.5 \approx 10$ μm in diameter. The thickness of the positive electrodes and the areal density of the sample are about 30 μm and 1 mg/cm^2, respectively. This positive electrode is coated on an Al current collector 25 μm thick. The positive electrode is charged by attaching it to a separator and a Li metal foil for the negative electrode. The charge current density is 45 mA/cm^2 (the C rate is 30), and the charging time is 1 min. More details about the sample preparation procedure and characterization are reported in Refs. [27, 28].

The beam spot and the current are selected to be 2×2 mm^2 and ~15 nA, respectively, for the measurements carried out at the National Accelerator Center, Seville University [29]. The result of the experiments is shown in Fig. 5.23, Chapter 5. Figure 5.23 shows a typical NRA spectrum for a LFP sample. The two edges observed in the high-energy region of the spectrum of α-particles between channel 600 and channel 920 correspond to the $^7Li(p,\alpha_0)4He$ and $^{19}F(p,\alpha_0)^{16}O$ nuclear reactions (see the inset in Fig. 5.23). F is not an active particle of the sample but is contained in the binder. By coincidence the energies of the emitted α-particles in both nuclear reactions are nearly equal: 7.709 MeV for ^7Li and 7.891 MeV for ^{19}F. Unfortunately, overlapping of these two peaks cannot be avoided by any experimental tool and deconvolution of the two peaks can be only done by an appropriate data analysis.

The distribution of the Li concentration as a function of depth in the sample is determined by comparing measured and calculated spectra as discussed in Fig. 5.24, Chapter 5, of this book. To calculate spectra for a given Li distribution, the commercial computer code simulation for nuclear reaction analysis (SIMNRA) is used [30]. The cross-section input related to the Li and F nuclear reactions were taken from Paneta et al. [31] and introduced as an R33 file in the SIMNRA simulation [21]. The Li distribution is determined as shown in Figs. 5.24b and 5.24d.

The actual procedure for determining the Li distribution is as follows. First the spectrum is calculated by assuming that the Li and F concentrations are homogeneously distributed along the whole depth. As illustrated in Fig. 5.24a, clear discrepancies are observed between the measured and calculated spectra. In particular, the measured yield at depths well below the sample surface is significantly higher than the calculated one. There are three possibilities for these discrepancies: (i) The Li concentration along the depth is not homogeneous, (ii) the F concentration along the depth is not homogeneous, and (iii) both are not homogeneous. However, since the distribution of the binder in the sample is homogeneous and F is contained in the binder, there is no physical reason to assume that the F distribution along the depth is inhomogeneous. Therefore, in order to properly fit the spectrum, only the Li distribution is assumed to be inhomogeneous along the electrode thickness (depth). To calculate the spectrum under this assumption, the sample is divided into several layers, each of them containing a different Li concentration. Then, simulations are carried out by varying the number of layers and their Li concentrations until a good fit is achieved. As depicted in Fig. 5.24b, the best fit is obtained by assuming that the total sample consists of 12 layers, each containing a certain amount of Li, which increases from 4.9 atomic percent (at%) for the layer close to the separator to 5.8 at% for the layer close to the current collector. The Li concentration in each layer is shown in Fig. 5.24c. These results indicate that the Li concentration increases with increasing depth from the separator.

The effects of the cover for shielding samples have been explored by using the LNO samples fabricated by the Toyota Central Research Laboratories. Figure 8.18 shows that the α-particles lose energy when they pass through the film cover. The α-energy spectrum without the cover has the upper edge at about 7.6 MeV and the lower edge at about 5.5 MeV, according to the α-spectrum of Fig. 8.18. However, in the case of Fig. 8.16, because of the pile-up problem, the lowest α-edge is not clearly seen. When there is a cover, the maximum α-particle energy decreases by 0.7 MeV for 7.5 μm^t, 1.1 MeV for 12.5 μm^t, 2.3 MeV for 25 μm^t Kapton films, and

2 MeV for a polymer-laminated Al cover. These results indicate that NRA characterization of the Li profile in the in situ measurement is applicable for a cover thinner than 12.5 μm^t of Kapton equivalent. Another issue is the reduction of the pile-up in order to store clean data for determining local depth distribution.

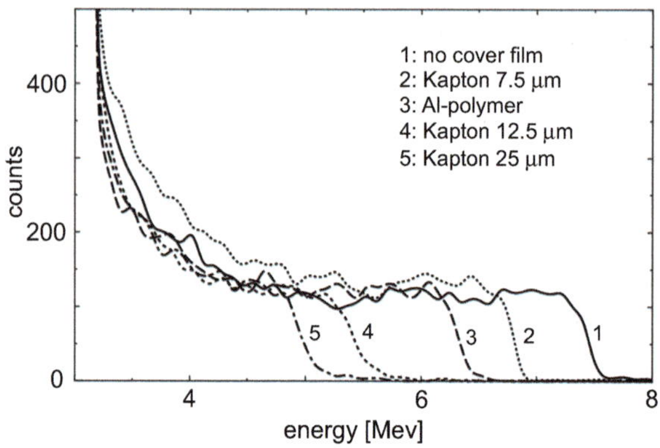

Figure 8.18 Proton and α-particle energy spectra. The curves 1, 2, 3, 4, and 5 are for the samples of no cover and those covered by 7.5 μm$^\tau$ Kapton, Al polymer laminate bag, 12.5 μm$^\tau$ Kapton, and 25 μm$^\tau$ Kapton, respectively. The signal at 3 MeV is the incident proton, and the high-energy tail from 3 MeV to 4.5 MeV is the pile-up signal due to a too-high flux to the detector. The high-energy-edge downward shift is due to the energy loss of α-particles in the cover.

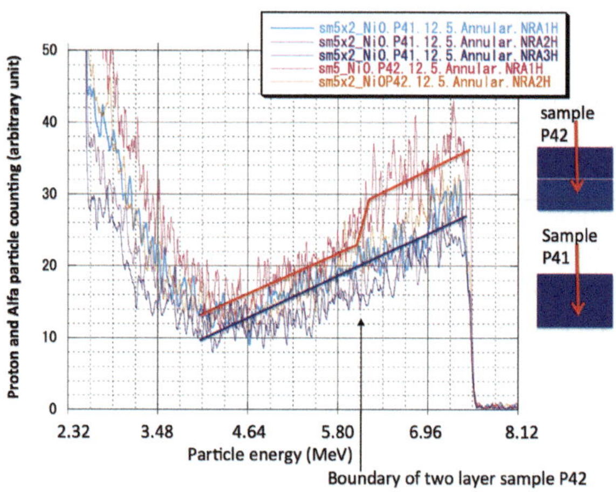

Figure 8.19 Energy spectra of proton and α-particles from the LIB's electrode irradiated by a 3 MeV proton beam. The RBS corresponds to the proton spectrum, and the NRA corresponds to the α-energy spectrum.

The α-spectra of red and yellow-red curves of Fig. 8.19 indicate that the Li concentration jumps by 25% at the interface of two layers of the P42 sample are observable. So, we conclude that Fig. 8.19 is the evidence of the proof of principle for applying NRA to the in situ measurement of Li depth distribution of LIB electrodes, where the samples are covered by 12.5 μm^{τ} Kapton.

8.6 Summary and Critical Issues

Various methods for in situ characterization of LIBs have been developed. For microscopic in situ diagnostics, X-ray diffraction and spectroscopy have been widely applied and matured. Neutron diffraction and scattering have also been developed recently and applied for the characterization. On the other hand, mesoscopic (spatial scale of submicrons ≈ millimeters) diagnostics are still under development. NDP and IBA are expected to be powerful tools for this purpose.

In situ IBA characterization of LIBs has been reviewed. A test cell for in situ measurements has been fabricated and irradiated by the proton microbeam at TIARA. Time-dependent variation of Li distributions in the positive electrode with a liquid electrolyte has been successfully identified with PIXE and PIGE. The application of this technique in ASSLIBs has also been tested. The application to ASSLIBs is still under investigation in collaboration with the Takasaki Institute of QST, the Graduate School for the Creation of New Photonics Industries, and the Jülich Research Center in Germany. The NRA of IBA is also applicable for profiling the Li depth distribution. Although in situ measurements have not yet been demonstrated, NRA seems to be a powerful tool for in situ measurements of the Li distribution in an electrode.

The critical issues in the application of IBA in in situ measurements are the following:

- Electrodes and the electrolyte are damaged by the ion beam.
- Required irradiation duration is relatively long in comparison with the time scales of high charge and discharge rates.
- The present IBA is not applicable to thick covered commercial LIBs because of the finite penetration depth of the ion beam.

Further development and testing of IBA techniques are expected for the in situ measurement of working LIBs.

References

1. Yang, Y. Q. and Yu, X. (2015). Advanced in situ diagnostic techniques for battery materials, http://energy.gov/sites/prod/files/2015/06/f23/es059_yang_2015_o.pdf
2. Abraham, D. P., Dees, D. W., Knuth, J., Reynolds, E., Gerald, R., Hyung, Y.- E., et al. (2005). Diagnostic examination of Generation 2 lithium-ion cells and assessment of performance degradation mechanisms, eds. Abraham., D. P., Oak Ridge National Laboratory, TN Report # ANL-05/21. Available from: http://www.osti.gov/bridge/product.biblio.jsp?query_id=0&page= 0&osti_id=861617&Row=0&formname=basicsearch.jsp
3. Williams D. B. and Carter C. B. (2004). *Transmission Electron Microscopy: Textbook for Materials Science*, Vol. 1, First ed. (Plenum, New York and London).
4. Nonaka, T., Okuda, C., Ukyo, Y. and Okamoto, T. J. (2001). In situ XAFS study on cathode materials for lithium-ion batteries, *Synchrotron Radiat.*, **8**, pp. 869–871.
5. Orikasa, Y., Koyama, Y., Uchimoto, Y., Ogumi, Z., et al. (2013). Transient phase change in two phase reaction between $LiFePO_4$ and $FePO_4$ under battery operation, *Chem. Mater.*, **25**, pp. 1032–1039.
6. Li, Z. G., Harlow, R. L., Gao, F., Lin, P., Miao, R. and Liang, L. (2003). Crystalline characterization of an operational Li-ion battery by synchrotron X-ray diffraction, *J. Electrochem. Soc.*, **150**, pp. A1171–A1175.
7. Binnig, G., Quate, C. F. and Gerber, C. (1986). Atomic force microscope, *Phys. Rev. Lett.*, **56**, pp. 930–933.
8. Tun, Z., Noël, J. J., Bohdanowicz, Th., Cao, L. R., Downing, R. G. and Goncharova, L. V. (2010). Cold-neutron depth profiling as a research tool for the study of surface oxides on metals, *Can. J. Phys.*, **88**, pp. 751–758.
9. Nagrapure, S. C., Downing, R. G., Bhushan, B., Babu, S. S. and Cao, L. (2011). Neutron depth profiling technique for studying aging in Li-ion batteries, *Electrochim. Acta*, **56**, pp. 4735–4743.
10. Whitney, S., Biegalski, S. R. F., Huang, Y. H. and Goodenough, J. B. (2009). Neutron depth profiling application to lithium-ion cell research, *J. Electrochem. Soc.*, **156**, pp. A886–A890.
11. Nagao, M., Pitteloud, C., Kamiyama, T., Otomo, T., Itoh, K., Fukunaga, T., Tatsumi, K. and Kanno, R. (2006). *Structure characterization and*

lithiation mechanism of nongraphitized carbon for lithium secondary batteries, J. Electrochem. Soc., **153**(5), pp. A914–A919.

12. Seidlmeyer, S., Hattendorff, J., Buchberger, I., Karge, L., Gasteiger, H. A. and Gilles, R. (2015). In operando small-angle neutron scattering (SANS) on Li-ion batteries, *J. Electrochem. Soc.*, **162**(2), pp. A3116–A3125.

13. Wang, H., Downing, R. G., Dura, J. A. and Hussey, D. S. (2012). In situ neutron techniques for studying lithium ion batteries, in *Polymers for Energy Storage and Delivery: Polyelectrolytes for Batteries and Fuel Cells*, Vol. 1096 (American Chemical Society), pp. 91–106.

14. Stancovski, V. and Badilescu, S. (2014). In situ Raman spectroscopic–electrochemical studies of lithium-ion battery materials a historical overview, *J. Appl. Electrochem.*, **44**, pp. 23–43.

15. Novak, P., Goers, D., Hardwick, L., Holzapfel, M., Scheifele, W., Ufheil, J. and Wursig, A. (2005). Advanced in situ characterization methods applied to carbonaceous materials, *J. Power Sources*, **146**, pp. 15–20.

16. (a) Ito, T. and Buturi, O. (2016). In situ Raman spectroscopy on the surface electrochemical studies, *Oyo Buturi*, **85**(9), pp. 754–760 (in Japanese). (b) Anzue, N., Ito, T., et al. (2003). In situ Raman spectroscopic study of thin-film $Li_{1-x}Mn_2O_4$ electrodes, *Solid State Ionics*, **156**, pp. 301–307.

17. Tadic, T., Jaksic, M., Capiglia, C., Saito, Y. and Mustarelli, P. (2000). External microbeam PIGE study of Li and F distribution in PVdF/HFP electrolyte gel polymer for lithium battery application, *Nucl. Instrum. Methods Phys. Res., Sect. B*, **161–163**, pp. 614–618.

18. Tadic, T. Jaksic, M., Medunioc, Z., Quarterone, E. and Mustarelli, P. (2001). Microbeam studies of gel-polymer interfaces with Li anode and spinel cathode for Li ion battery applications using PIGE and PIXE spectroscopy, *Nucl. Instrum. Methods Phys. Res., Sect. B*, **181**, pp. 404–407.

19. Mima, K., Gonzalez-Arrabal, R., Azuma, H., Yamazaki, A., Okuda, C., Ukyo, Y., Sawada, H., Fujita, K., Kato, Y., Perlado, J. M. and Nakai, S. (2012). Li distribution characterization in Li-ion batteries positive electrodes containing $Li_xNi_{0.8}\,Co_{0.15}Al_{0.05}\,O_{0.2}$ secondary particles (0.75 6 x 6 1.0), *Nucl. Instrum. Methods Phys. Res., Sect. B*, **290**, pp. 79–84.

20. Habrioux, A., Surble, S., Berger, P., Khodja, H., D'Affroux, A., Mailley, S., Gutel, T. and Patoux, S. (2012). Nuclear microanalysis of lithium dispersion in $LiFePO_4$ based cathode materials for Li-ion batteries Author links open overlay panel, *Nucl. Instrum. Methods Phys. Res., Sect. B*, **290**, pp. 13–18.

21. Gonzalez Arrabal, R., et al. (2015). Meso-scale characterization of lithium distribution in lithium-ion batteries using ion beam analysis techniques, *J. Power Sources*, **299**, pp. 587–595.

22. Yamazaki, A., Orikasa, Y., Chen, K., Uchimoto, Y., Kamiya, T., Koka, M., Satoh, T., Mima, K., Kto, Y. and Fujita, K. (2016). In-situ measurement of the lithium distribution in Li-ion batteries using micro-IBA techniques, *22nd International Conference on Ion Beam Analysis (IBA 2015)*, Vol. 371, pp. 298–302.

23. Kamiya, T., Suda, T. and Tanaka, R. (1996). High energy single ion hit system combined with heavy ion microbeam apparatus, *Nucl. Instrum. Methods Phys. Res., Sect. B*, **118**, p. 447.

24. Sakai, T., Kamiya, T., Oikawa, M., Satoh, T., Tanaka, A. and Ishii, K. (2002). JAERI Takasaki in-air micro-PIXE system for various applications, *Nucl. Instrum. Methods Phys. Res., Sect. B*, **190**, pp. 271–275.

25. Ziegler, J. F., Biersack, J. P. and Ziegler, M. D. (2013). SRIM-2013, The stopping and range of ions in matter, http://www.srim.org.

26. Sagara, A., Kamada, K. and Yamaguchi, S. (1988). Depth profiling of lithium by use of the nuclear reaction 7Li(p, α)4He, *Nucl. Instrum. Methods Phys. Res., Sect. B*, **34**, pp. 465–469.

27. Orikasa, Y., Maeda, T., Koyama, Y., Murayama, H., Fukuda, K., Tanida, H., Arai, H., Matsubara, E., Uchimoto, Y. and Ogumi, Z. (2013). Direct observation of a metastable crystal phase of Li_xFePO_4 under electrochemical phase transition, *J. Am. Chem. Soc.*, **135**, pp. 5497–5500.

28. Orikasa, Y., Maeda, T., Koyama, T., Murayama, H., Fukuda, T., Tanida, H., Arai, H., Matsubara, E., Uchimoto, Y. and Ogumi, Z. (2013). Transient phase change in two phase reaction between $LiFePO_4$ and $FePO_4$ under battery operation, *Chem. Mater.*, **25**, pp. 1032–1039.

29. Garcıa Lopez, J., Ager, F. J., Barbadillo Rank, M., Madrigal, F. J., Ontalba, M. A., Respaldiza, M. A. and Ynsa, M. D. (2000). CNA: the first accelerator-based IBA facility in Spain, *Nucl. Instrum. Methods Phys. Res., Sect. B*, **161–163**, pp. 1137–1142.

30. Mayer, M. (1997). SIMNRA user's guide, Max-Planck-Institut *für* Plasmaphysik, Garching, Germany, n.d. IPP 9/113.

31. Paneta, V., Kafkarkou, A., Kokkoris, M. and Lagoyannis, A. (2012). Differential cross-section measurements for the 7Li(p,p0)7Li, 7Li(p,p1)7Li, 7Li(p,α0)4He, 19F(p,p0)19F, 19F(p,α0)16O and 19F(p,α1,2)16O reactions, *Nucl. Instrum. Methods Phys. Res., Sect. B*, **288**, pp. 53–59.

Index

For Product Safety Concerns and Information please contact our EU representative GPSR@taylorandfrancis.com Taylor & Francis Verlag GmbH, Kaufingerstraße 24, 80331 München, Germany

Batch number: 10726509

Printed by Printforce, the Netherlands